Seattle Public Library

HARD-ROCK MINERS

Hard-Rock
Miners

THE INTERMOUNTAIN WEST,
1860–1920

BY

Ronald C. Brown

Texas A&M University Press

COLLEGE STATION AND LONDON

Library of Congress Cataloging in Publication Data

Brown, Ronald C   1945–
   Hard-rock miners.

   Bibliography: p.
   Includes index.
   1. Miners—The West—History. I. Title.
HD8039.M61U45   331.7′62′2340978   78-21778
ISBN 0-89096-066-6

Manufactured in the United States of America
FIRST EDITION

To Judy

Contents

List of Illustrations   ix

List of Abbreviations   xi

Preface   xiii

Acknowledgments   xv

CHAPTER

  1. Meandering Miners and Majestic Mountains   3

  2. The Miners' Towns and Families   20

  3. The Perils and Pleasures of the Miners' World   36

  4. At Work amidst the Ore   59

  5. The Deadly Depths   75

  6. Earning a Living by Mining   99

  7. Miners and Their Managers   118

  8. The Need to Organize: The Miner and His Union   145

  9. "Only a Man in Overalls, Lay Him Anywhere—"   161

APPENDIX

  A. "Present-Day Mining"   169

  B. Accidents and Injuries Typical in Hard-Rock Mining in the
      Intermountain West Prior to 1920   173

Bibliography   177

Index   191

List of Illustrations

George Whitwell Parsons, early miner, 1880's   *following page 110*

Cripple Creek, Colorado, 1891

Bingham Canyon, Utah, showing natural environment of mining

The Comstock, showing urban environment of mining, 1876

Miners' cabins at Chicago Creek, Colorado

Burro train in Telluride, Colorado

One-room school in Camp Bird, Colorado, 1903

Boarding-house dining room at Bingham Canyon, Utah

Early twentieth-century boarding house, Bingham Canyon

Snowslide damage at Camp Bird Mine, 1906

Silver Plume, Colorado, baseball team, 1889

Miners playing keno

Rock-drilling contest in Goldfield, Nevada, circa 1905

Assay office at Utah's Ontario Mine

Ibex Mine in Leadville, a turn-of-the-century extractory

Freight wagon at Utah copper mine

Steam shovels at open-pit mine, Bingham Canyon, Utah

Colorado miners double-jack into stopes

Underground miners working in teams

Miners preparing to drill

Miners with mechanized equipment

Miner descending into Combination Shaft at Goldfield, Nevada

Horse-powered winch near Leadville, Colorado

Extractory cage system at Utah's Ontario Mine

Headframe and hoisting system at Congress, Arizona

Hoist-house equipment at Utah's Buckeye Mine
Union poster from Cripple Creek District, 1904
Smuggler Union miners at Pandora, Colorado, early 1900's
Only a man in overalls, a miner about to descend

List of Abbreviations

BL	Bancroft Library, University of California, Berkeley.
HL	Huntington Library, San Marino, California.
RCNM	State Records Center of New Mexico, Santa Fe.
UNM	Special Collection Department, Zimmerman Library of University of New Mexico, Albuquerque.
UA	Special Collections Department, University of Arizona Library, Tucson.
SHSC	State Historical Society of Colorado, Denver.
CU	Western History Collection, University of Colorado Libraries, Boulder.
DPL	Western History Department, Denver Public Library.
WHRC	Western History Research Center, University of Wyoming, Laramie.

Preface

ANY student of the popular mythology of the American West is certain to confront the stereotyped prospector. The fabled "sourdough" of bearded countenance and rumpled garb, accompanied always by his faithful, though cantankerous, burro, has long moved through the pages of history as the characteristic miner. Though he was indeed a miner, he rarely stayed to help build the mining towns. Instead, he was replaced by less colorful but more interesting and imaginative men who proudly called themselves hard-rock miners and who labored in the nonferrous mineral mines, especially those of gold, silver, lead, and copper. In part, these hard-rock stiffs have been ignored because they blazed few trails and were credited with few new strikes. They were predominantly diggers, blasters, builders, and destroyers rather than pathfinders.

Mining itself was a masculine vocation whose practitioners came from around the world and throughout the United States. Once they reached the western mining frontier, they continued to move about, searching constantly for the fulfillment of individual dreams. Additionally, the nature of mineral deposits and the procedures of extraction meant that few miners enjoyed a steady job in a stable community. They found work in hundreds of precious and nonferrous base metal camps throughout the Trans-Mississippi West. Some men married and raised frontier families, and all wrestled with the environment and the problems of urban living. Their work was dangerous and exacting, and the boss was often unappreciative. When the indifference of profit-oriented corporations convinced miners that they had to unite and to cooperate for their own protection, they formed unions. The hard work of extracting ore from mines took its toll on the health and lives of these men.

This study focuses primarily on the area now encompassed by the states of Wyoming, Colorado, New Mexico, Arizona, Utah, and

Nevada. This vast expanse, which incorporates some of the most important mining districts of the nineteenth and early twentieth centuries, also provides a wide sampling of the varied environments and geological formations that Western miners encountered. Covering the years between 1859 and 1920, the work follows these men through the rushes into an era when mining had become increasingly mechanized and the earlier generation's unique skills unnecessary. Hardrock mining remained one of the nation's most dangerous occupations and required a distinctive temperament, but the pride and sense of identity began to pass.

Unlike other students of the period, I do not believe that studies of the mining frontier need conclude at some specific time chosen to reflect the culmination of the process called industrialization. By any standard available, it is certain that the mining industry underwent continuous industrialization throughout this period. By the late 1860's and 1870's mines in communities like Virginia City, Nevada, and Central City, Colorado, had become industrial "extractories" that employed hundreds of men, adopted the latest technological innovations and steam-powered machinery, implemented occupational specialization, and depended upon investment capital and trained managers. Other mines, however, lagged behind in the industrialization process. Mining was an undertaking in which the stage of development of a given mine was more closely related to the time interval since its discovery than to the state of industrialization of the occupation as a whole.

Acknowledgments

As this study has gradually taken shape, I have incurred debts of gratitude to numerous institutions and individuals who have made my work both more enjoyable and easier. In particular I should like to thank the staffs of the University of Illinois Library, the Library of Southwest Texas State University, the Bancroft Library, the Huntington Library, the Records Center of New Mexico, the Special Collections Departments of the University of New Mexico's Zimmerman Library and the University of Arizona Library, the State Historical Society of Colorado, the Western History Department of Denver Public Library, the Arizona Historical Society, and the Utah State Historical Society. Special thanks are due Gene Gressley of the Western History Research Center, University of Wyoming, and John Brennan of the Western Historical Collections, University of Colorado. Both took time to assist me in winnowing masses of archival data, while explaining both the limitations and possibilities of manuscript research. I also wish to thank Albert J. Frantz, former official of the United Steelworkers of America, District 38, who discussed with me his activities as a miner and union member. In addition, I wish to thank Mr. Ray M. Marro for drafting the illustrations of mines.

Since 1975 my colleagues at Southwest Texas State University have provided encouragement that has kept me plodding forward. Specifically, James A. Wilson has taken time from a busy schedule to counsel me on the mechanics of preparing a manuscript for publication. For the support of three good friends and teachers, James J. Barnes, Clark C. Spence, and Duane A. Smith, I shall be ever grateful. Finally, I thank my wife, Judy, for her patience, prodding, and unfaltering encouragement.

HARD-ROCK MINERS

1.

Meandering Miners and
Majestic Mountains

In the mid-nineteenth century, men first began to prospect the inter-mountain West (the region between the eastern slope of the Rockies and the Sierra Nevadas). Earlier gold rushes in Georgia and California convinced the far-sighted and optimistic that the Rocky Mountains and Great Basin surely contained untold quantities of precious and base metals. This enthusiasm fostered the nearly simultaneous discoveries of silver on the Comstock Lode and gold near Pike's Peak. In 1859 the western and eastern perimeters of this vast hinterland demonstrated their potential mineral wealth. For the next half-century men searched remote parts of the region for other bonanzas. They were intermittently rewarded, and American society adjusted to the phenomenon of "rushes." Individuals dropped whatever they were doing and joined the hordes of seekers in search of their "pots of gold." As every would-be prospector and miner soon learned, western streets were not paved in gold, nor were nuggets lying about for the taking. Instead, these immigrants encountered an inhospitable climate and all the human problems that they had believed behind them. Many came west, looked hard about them, and decided that they would be better off if they returned posthaste to the States. Others stayed and worked in the mines; it is their story that will be told. The environment they encountered, the conditions under which they labored, and the lives they built for themselves and their families— these are the matters to be explored.

Few of those who came west in the early days expected to spend their lives toiling in the stopes of another man's mine; rather, like Virginia City's James Galloway, they expected to get rich quick. Thomas D. Sanders later described how the Pike's Peak fever intoxicated his father, who promptly settled his affairs and took his family west. "It reached the slumbering little town of Girard [Illinois] and Father. A wonderful transformation took place in him. The effect was electrical. He forgot his worn out condition and hard work and became all fired up again with the ambitions of his earlier years."[1] Like the elder Sanders or George A. Carpenter, who followed the gold fever to South Pass, Wyoming, in 1870–1871, westbound men deserted stable societies, expecting to make a fortune in the territories. Thirty years later their descendants joined rushes to the Nevada camps of Tonopah, Goldfield, and Rawhide, thereby demonstrating that the dream of sudden wealth outlived the era when men knew little about the intermountain West.[2]

Not all came west with expectations of making strikes themselves; many turned to hard-rock mining because wages were two to five times greater than those paid industrial and agricultural workers. George W. Parsons carefully weighed his prospects before leaving San Francisco for Tombstone in 1880. On January 5, he concluded, "I think after all that the mines promise more for me and I wouldn't be surprised to find myself in them before many days." Eight days later he was "half determined . . . to go to the Tombstone district and begin as a common miner." He professed to want to "learn the business and stick to it." The prevailing three- to five-dollar daily wage attracted the strong and the socially rootless. They reasoned that a hard-working, frugal man could save his pay and build a nest egg with which to return home or to start anew.[3]

Experienced miners understood the dangers of their vocation, but they noted the extraordinary wage differentials between western mines and those elsewhere. Whether working in Pennsylvania, Cornwall, Ireland, or Italy, they saw new opportunities in the new fields. Joining the procession of individuals seeking a better future in the American West, they mingled with farmers, bookkeepers, common laborers, and professional men in the stopes of hard-rock mines.

[1] John Debo Galloway, *Early Engineering Works Contributory to the Comstock*, p. 15; Thomas D. Sanders and Charles C. Stemmer, "Hardships and Perils on the Western Frontier or Fifty Years of Pioneering," pp. 16, 39–40, 44.

[2] "Heyday of Wyoming Gold Mining Camps Recalled by Pioneer: George A. Carpenter . . . ," newspaper clipping of unknown origin, Mining File, Western History Research Center, University of Wyoming, Laramie (hereafter cited as WHRC).

[3] George Whitwell Parsons, *The Private Journal of George Whitwell Parsons*, January 5 and 13, 1880, pp. 70, 75.

Some of the European miners encouraged their countrymen to follow them across the Atlantic and join the throng. The Cornish miners even acquired the sobriquet "Cousin Jacks," reputedly because they always claimed to have a "cousin" back in Cornwall who had the very skills an employer sought. Traditionally clannish, Cornishmen like Richard Harvey of Caribou, Colorado, worried about their friends and relatives still struggling to survive back home. "I get a letter occasionally from father and George. Poor old man how I do pity him. He certainly has not had an envious life since Mother's death. He and George batching in our house together with no one to do anything for them—and after working all day trying to make a little bread."[4] Similarly, Greeks and other southeastern Europeans, immigrating in the years after 1890, sent money home and expounded the virtues of the frontier wage scale.

Health concerns brought some men west. Nineteenth-century physicians prescribed mountain air for assorted respiratory ailments, especially tuberculosis, whose victims occasionally found their way into the mines. Only a later generation, aware of the dangers from dust, would discourage tubercular men from seeking employment in the mines. Not only their own health problems but also the illnesses of their sons and daughters brought men to the western states. Once there, some turned to mining to sustain themselves and their families. One, Coloradan Henry Bouyer, appealed for work, admitting that he knew little of mining but assuring a prospective employer that he "could very soon learn."[5]

Men also came west in response to the deliberate inducements of railroad and mining corporations. Seeking the passengers and settlers essential to their economic growth, the railroads encouraged men to join the rush to the mines. They publicized opportunities and promoted migration with special rates and benefits for those willing to settle in the West. Persistent labor shortages persuaded mining companies to advertise both for skilled miners and for common laborers.

[4] R. Harvey to M. Duke McCallister, August 21, 1874, in "Old Caribou and Central City: Glimpses from the Letter Book of Cornishman Richard Harvey," Agnes Wright Spring Papers, WHRC.

[5] Henry Bouyer to John F. Campion, Salida, Colorado, May 17 and May 24, 1894, John F. Campion Papers, Western History Collection, University of Colorado Libraries, Boulder (hereafter cited as CU).

Lured by generous promises, recruits occasionally discovered that company officials had deceived them or intended to use them as strikebreakers. Whatever the motives, corporations successfully promoted the migration of men who otherwise would not have become miners.[6]

The "rushers," the relocating miners, and those simply seeking a better standard of living all shared an initial optimism and a conscious decision to emigrate. Other men, however, fled west not so much to search for a goal as to escape from a past. Running from an unpleasant or sometimes notorious existence, these fugitives were searching for anonymity in an industrial occupation. Escaping personal commitments or social crimes, most of them kept to themselves and drifted from camp to camp, leaving few records. Like pimps, prostitutes, and gamblers, most of them remained anonymous, living, working, and dying in the remote camps of the mining West. They learned that prospective miners needed only provide an affirmative answer to the simple question, "Can you do the work?" Young Frank Crampton recalled the events that led him into mining and subsequently mining engineering: "My practical education started on a November day when, as a lad of sixteen, busted from an old ivy league college, broke after running away from home and a family that thought I had disgraced it forever, I was taken in tow by two hard-rock mining stiffs and shown the ropes."[7] This chance encounter reshaped his entire life. Likewise, J. J. Johnson, a leader of Cripple Creek miners during the labor strike of 1893, turned to mining after he had been dismissed from West Point. Men like Crampton and Johnson found mining a career in which their past transgressions were ignored.[8]

As labor strikes became more common in both the east and the west, some men discovered that emigration was imperative. Fearing either physical reprisals or blacklisting, they moved from troubled

[6] General Ticket and Passenger Department of the Chicago, Rock Island and Pacific Railroad, *Colorado and Its Mines* 1 (May, 1880), Mining File, WHRC; "The Oro City Affair," *Georgetown* (Colorado) *Courier*, May 24, 1877.

[7] Frank A. Crampton, *Deep Enough: A Working Stiff in the Western Mine Camps*, pp. 30–31.

[8] Vernon H. Jensen, *Heritage of Conflict: Labor Relations in the Non-Ferrous Metal Industry up to 1930*, p. 45.

communities to peaceful ones. Others moved simply because they wanted to work but could not bring themselves to cross the picket lines. Whatever their precise reasons, these men became another current in the stream of men who turned to mining elsewhere as a means of escape.[9]

Then, too, there were the social and psychological casualties of the Civil War. In 1864, Nathaniel P. Hill, principal organizer of Colorado's early smelting industry, reported meeting many "emigrants . . . from Missouri and Arkansas," who had fled as guerrilla warfare threatened the remnants of the noncombatant population. Hill and his associates stopped to converse with them. "We frequently asked them whither they were bound, and were generally told they did not know." Some of these migrants resettled in agricultural communities or in the growing commercial centers of the American West, but others trickled into the mining camps to find work. Not only civilians but also veterans drifted westward. Accustomed to the uncertainties and attendant excitement of conflict, men like Johnson Barbee, Jacob Waldorf, and Carlyle C. Davis turned westward and found in mining a psychological surrogate for the thrill and camaraderie of the campaign.[10]

Others who came west were instinctive drifters who traveled about the country. Alfred P. Camp recorded the exploits of two such men whom he encountered in 1875, as they sat about a campfire in the San Juan Mountains. They had spent earlier years at sea and had passed their lives wandering from place to place and from job to job.[11]

As men came west for varied reasons, so they came in varied states of preparedness. Some arrived with nothing more than the clothes they wore, while others carried prized or needed possessions.

[9] Ibid., pp. 219–220; Marshall Sprague, *Money Mountain: The Story of Cripple Creek Gold*, pp. 135–136; Robert Livermore, "An Autobiography," p. 90, Robert Livermore Papers, WHRC.

[10] Nathaniel P. Hill to Wife, June 14, 1864, Mrs. Hill's transcription, Nathaniel P. Hill Papers, State Historical Society of Colorado, Denver (hereafter cited as SHSC); Mabel Barbee Lee, *Cripple Creek Days*, pp. 1–5; John Taylor Waldorf, *A Kid on the Comstock: Reminiscences of a Virginia City Childhood*, ed. Dolores Bryant Waldorf, pp. 4–7; Carlyle C. Davis, *Olden Times in Colorado*, pp. 33–83.

[11] Journal of Alfred P. Camp, Alfred P. Camp Papers, July 27, 1875, SHSC.

The well-prepared and financially secure transported their own bedding, staples, utensils, tools, clothing, and miscellaneous memorabilia so that they could create a new home in the mining camps. George Parsons enumerated his acquisitions and expenditures:

Packed today and tonight. Outfit consists of 2 pairs of double blankets— 4 prints $13.00—one rubber blanket $2.00—rubber pillow $3.25—rubber leggings $1.25—rubber cap (soft) 35¢—cork screw 35¢—miners shirts (3) $4.50—riveted overalls $1.85—dust cover for pillow 75¢ and sundries—F. shirts—drawers—W C paper—tooth brush—paper collars and cuffs—bone studs—trunk strap—blacking—stationery and clothes repaired $6.75, making total of $34.05. Ticket through to Tucson $57.50— Grand total $91.55—$7 to Tombstone and necessary expenses at Tucson will make amt over $100—considerably.

Not only did he detail his clothing and supplies before he set out, but he also talked with friends and acquaintances about life in Arizona. He read books about mining and about Arizona and initiated a physical fitness program designed to increase his strength and physical stamina. Few men left for the mines with more systematic preparation than Parsons; however, even he still had to learn the mechanics of mining, and he discovered that no amount of study could have prepared him for the rigors he encountered.[12]

The population of the mining frontier was disproportionately foreign in origin, in part because Americans had had only scattered contact with hard-rock mining before the California gold rush. In the 1870 census, between 40 and 50 percent of the Nevada White Pine District's population was non-American, and in 1880, fully one-third of Leadville's nearly fifteen thousand people were immigrants. According to the Virginia City census of 1880, native-born Americans accounted for less than 28 percent of the mining work force, whereas Irishmen made up 30 percent and Englishmen (including the Cornish) another 23 percent. These percentages reflect the relative importance of immigrants in the mining camps at a time when they represented only 13 percent of the national population and a slightly higher 23 percent of the western settlers. During the nineteenth century the British Isles, especially Cornwall and Ireland, provided the largest body of immigrant miners. Brought to America by domestic depres-

---

[12] Parsons, *Journal*, September 9 and 11, 1879, January 5, 12, 15, 16, 17, 20, 21, 22, 31, and February 3, 1880, pp. 34–35, 70, 74, 75–81, 82.

sions, they dominated the scene until displaced in the late nineties by central and southeastern Europeans.[13]

The influx of these new immigrants had become so significant by 1900 that the Western Federation of Miners (WFM) considered and adopted provisions requiring the translation of union documents and rituals into Italian, Slavic, and Finnish. These immigrants brought distinctive cultural traditions, which colored community lifestyles. However, Orientals, Spanish-speaking Americans, and Indians faced severe discrimination. Routinely denied jobs in the mines, the Chinese operated laundries, restaurants, and dens pandering drugs and prostitution. While Mexican-Americans remained an important component of the Southwest's mining labor force, they endured persistent inequities even there and in other regions were excluded from underground work altogether. Though present in the mining communities, American Indians were accorded the lowest status, subjected to ridicule, and relegated to scavenging for their survival.[14]

Having become miners, men discovered that their lives were unpredictable and frequently migratory. Mining ventures ordinarily persisted only as long as mineral ore merited the costs of extraction, applied technology kept pace with the needs of the excavation process, and capital was available to keep the operation running. Duane A. Smith reported a classic case of geographical mobility in the career

[13] No thoroughgoing statistical investigation of the character of mining communities' populations is available. Duane A. Smith's "The San Juaner: A Computerized Portrait," *Colorado Magazine* 52 (Spring, 1975), 137–152, is a singular attempt to apply quantification to the intermountain frontier. Using samples from the federal censuses of 1860, 1870, and 1880 plus the Colorado census of 1885, he finds that foreigners made up about 32 percent of the mining population but a substantially smaller portion of the total population. The figures cited in the text are drawn from W. Turrentine Jackson, *Treasure Hill: Portrait of a Silver Mining Camp*, pp. 27, 204; Rodman W. Paul, *Mining Frontiers of the Far West, 1848–1880*, p. 28; Eliot Lord, *Comstock Mining and Miners*, p. 383; Bureau of Census and Social Science Research Council, *The Statistical History of the United States from Colonial Times to the Present*, pp. 11–12.

[14] *Official Proceedings of the Tenth Annual Convention of the Western Federation of Miners of America, Denver, Colorado, May 26–June 7, 1902* (Denver: Colorado Chronicle, 1902), p. 131; *Official Proceedings of the Eleventh Annual Convention of the Western Federation of Miners of America, Denver, Colorado, May 25–June 10, 1903* (Denver: Western Newspaper Union, 1903), p. 123.

of Coloradan Richard Irwin. In the twenty years after 1860, Irwin lived and worked in Central City, Buckskin Joe, Tarryall, Georgia Gulch, Gold Run, Empire, Georgetown, Rosita, the San Juans, and the Gunnison country. He also had made excursions into New Mexico, Utah, and South Dakota. Even in boom periods would-be miners discovered that jobs were not as plentiful as they had anticipated. In 1861, George O. James, another Coloradan, glumly reported, "I never thought that if I came down to hard labor I could not get even that to do but so it is."[15]

Circumstantial evidence suggests that the geographic mobility of miners was linked not only to occupational opportunities, but also to occasional wanderlust. For example, the story of James Galloway suggests that he moved to new places, even when there was no specific need. Cripple Creeker Ross B. Moudy had simply wandered in from Florence, Colorado, although his subsequent move to Wyoming stemmed from labor difficulties that embroiled Cripple Creek and the surrounding communities. Others, like George Struber of Leadville, worked steadily in one vicinity for some time and then simply disappeared.[16]

Often men even abandoned mining temporarily and experimented with other occupations. Erstwhile miners dabbled in shopkeeping, farming, ranching, lumbering, teaching, freighting, and railroading. Others turned to the livery business or to veterinary medicine; a few became local constables and politicians. Joe Hill, the Wobbly bard and martyred folk hero, provides an interesting example of the nomadic and occupationally varied lifestyle of the group. Between 1902 and 1914 Hill, a Swedish immigrant, worked as a copper miner, a seaman, a migratory farm laborer, and a pipe layer. This pattern of movement was common, as evidenced by the reports of Arizona's Phelps Dodge Company, which claimed more than 100 percent annual labor turnover into the 1920's.[17]

[15] Duane A. Smith, "Colorado's Urban-Mining Safety Valve," *Colorado Magazine* 48 (Fall, 1971), 307; Diary of George O. James, January 3, 1861, Western History Department, Denver Public Library (hereafter cited as DPL).

[16] Galloway, *Engineering Works*, pp. 13, 19; Ross B. Moudy, "The Story of a Cripple Creek Miner," *Independent* 57 (August 18, 1904), 380; S. W. Mudd to C. C. Chase, Leadville, March 28, 1901, and Mudd to C. N. Eaton, Leadville, May 4, 1901, Campion Papers, CU.

[17] Melvyn Dobofsky, *We Shall Be All: A History of the Industrial Work-*

Mining required youth and vigor; the work itself was strenuous, and the conditions of labor were rugged. The typical workers were therefore young bachelors who had few possessions and who developed few tangible attachments to their communities. Unburdened by conventional ties, they had no compunction about pulling up stakes and moving on. One group of seasonal or "snow-bird" miners routinely migrated to the northern mines in the summer and back to those of Utah, Arizona, and Nevada in the winter. Frank Crampton and his tutors—Sully, John T., and Jack—chose mining districts with an eye to the prevailing excitement, the climate, the presence of comrades, and the job prospects. During his first five years in the West, Crampton worked in Colorado, Nevada, and Utah. Caring only about the present and the future, such men ignored friends and relatives, even wives.[18]

Although they made important contributions to the economic and industrial development of the intermountain region, miners never totaled more than fifty thousand in the present states of Arizona, Colorado, Nevada, New Mexico, Utah, and Wyoming (see Table 1.).

Comparing the numbers of miners in the region over the years studied is difficult due to historical circumstances and irregularities in census procedures. First, inasmuch as the census of 1860 and the rushes to Nevada and Colorado occurred simultaneously, statistics for that year are unreliable. Second, in 1870 and 1880, but not thereafter, census takers did not distinguish between coal miners and metal miners. That differentiation, begun in 1890, accounts for the apparent decline in the number of miners between 1880 and 1890. Finally, comparisons over time can begin only in 1880, since the census of 1870 did not provide a territorial breakdown of occupational groupings. Thus the total number of miners for 1870 includes the miners of Montana, Idaho, and Washington with those of the other territories.

---

ers of the World, pp. 307–308; E. D. Gardner and D. J. Parker, "Safety Organizations in Arizona Copper Mines," U.S. Bureau of Mines, Technical Papers 452 (1929), 24. See also Interviews of the Colorado Pioneers (Amos Augustine Frost, Harry T. Cook, Dominick Faussone, Samuel S. Boucher, Charles Henry Fosdick, James A. Beatty, Joseph Dennis, George Heavrin, Henry H. Hammond, Thomas J. Doyle, Frank L. Wilson, John Griffith, Mrs. Dora Gout Moody, and William Orson Cairns), SHSC.

[18] Crampton, Deep Enough, pp. 32–58. See also variously titled notice and information sections of Miners' Magazine, 1900–1918.

TABLE 1

Number of Miners in the Intermountain West, 1870–1920

Location	1870	1880	1890	1900	1910	1920
Arizona		4,678	3,561	7,414	14,623	14,466
Colorado		28,970	16,476	22,793	13,406	8,208
Nevada	8,241*	6,647	4,844	2,637	9,024	5,343
New Mexico		1,496	2,168	3,259	2,598	3,187
Utah		2,592	3,120	5,624	6,810	5,081
Wyoming		328	1,151	1,232	632	169
TOTAL	25,746**	44,711	31,320	42,959	47,093	36,454

SOURCE: U.S. Department of the Interior and Bureau of the Census, Census Reports, Ninth through Fourteenth, 1870–1920, tables for occupations.

* In 1870, separate figures were reported for Nevada only because only Nevada was a state. The Census Office reported only a total for the others, all of which were territories.

** The 1870 total was reported for all territories. Thus it includes the miners of Montana, Idaho, and Washington with those of the territories on this table. Other totals reported here are for only the states and territories listed.

Nonetheless, the available figures indicate that Arizona, Colorado, and Nevada had the region's largest mining populations in the nineteenth century. By 1900 Utah had more miners than Nevada, but the Goldfield-Tonopah boom and the copper mines around Ely propelled Nevada back into the lead by 1910. However, these aggregate totals do not reflect the comparative importance of mining within the states and territories. Table 2 shows the mining population as a percentage of the gainfully employed. In New Mexico, Utah, and Wyoming miners never exceeded 7 percent of the working population; however, even here (except in Wyoming in 1920), they remained more important than in the nation as a whole.

Moving into the mineral districts of the Great Basin and the lower Rockies, miners confronted the vagaries of an inhospitable climate and an intractable terrain. Almost all the region exceeds one thousand feet in elevation. Most mining districts and towns were located in mountainous or semimountainous areas, and this high-altitude environment posed problems for newcomers. This area is also noted for its extreme temperatures; winters average 10° to 30° F, while summers range from 60° to 100° F. The average annual pre-

TABLE 2

Miners as a Percentage of the Gainfully Employed in the
Intermountain West, 1870–1920

Location	1870	1880	1890	1900	1910	1920
Arizona		21.0	13.5	13.9	16.6	11.1
Colorado		28.6	8.6	10.4	4.0	2.2
Nevada	30.6*	20.6	20.7	13.3	20.1	14.2
New Mexico		3.7	4.0	4.9	2.1	2.6
Utah		6.5	4.7	6.6	5.2	3.4
Wyoming		3.7	3.8	2.8	0.9	0.2
U.S.A.	1.2	1.3	0.6	0.7	0.4	0.3

SOURCE: U.S. Department of the Interior and Bureau of the Census, Census
Reports, Ninth through Fourteenth, 1870–1920, tables for occupations.

* In 1870, separate figures are reported for Nevada only because only
Nevada was a state. The others, all territories, were not reported by the Census
Office.

cipitation is the lowest in the continental United States, with some
localities receiving less than ten inches of rain per year.[19]

Miners and their associates had varied reactions to this environment. While complaints were more common than compliments, some
miners recorded favorable impressions of the weather and the climate.
Virginia City's James Galloway noted clear winter days, "lovely
moonlight nights," and "beautiful moonlight now." Utah towns also
elicited favorable comments. Park City was said to be a good camp
with its clean, scenic countryside and its proximity to Salt Lake City.
Writing from the Star District in 1870 and 1871, John Shumer had
nothing but praise for the region and its temperate climate. In late
March, 1871, he wrote, "This has been the mildest winter I have ever
spent anywhere." While admitting that there might be some disagree-
able weather ahead, he predicted "delightful" weather after April.[20]

The climate of south-central Wyoming constantly inspired one

[19] W. Eugene Hollon, *The Great American Desert: Then and Now*, p. 10;
Joseph E. Williams, ed., *The Prentice-Hall World Atlas*, pp. 38–43, 45.

[20] Galloway, *Engineering Works*, p. 17; Jensen, *Heritage of Conflict*, p.
259; John Shumer to J. Wilbur, Star District, Utah, December 6, 1870, and
March 31, 1871, Utah Mining Papers, Huntington Library, San Marino, California (hereafter cited as HL).

otherwise frustrated promoter and part-time miner of the Grand Encampment region. Major O. S. Alers' diaries reflect continual enthusiasm for the "pleasant," "fine," or "splendid" weather. He enjoyed the natural splendor of the Wyoming summer and frequently noted the abundant wild flowers, the wild berries, and the cool days and evenings.[21]

Colorado's climate elicited more complaints than other regions', especially during its seemingly interminable winters and the parts of spring, which seemed more like winter. However, Richard A. Pomeroy wrote this glowing account of Georgetown's winter of 1877: "I have never passed a more delightful Winter as to climate. . . . More than half the people here never wear overcoats unless they go on the mountain." While less enthusiastic, Cornishman Richard Harvey found Caribou, Colorado, a "very pretty place . . . in the summer."[22]

The winters of the Southwest were and are one of that region's greatest attractions. W. F. Staunton described them as delightful. "I remember one [year] when there was not a single frost all winter long." Staunton's praise was echoed by George Parsons: "The most delightful kind of a climate. Beats Florida."[23]

The pleasures of the Arizona winters were countered by the intense heat of the Arizona summers. Even Staunton admitted that summers in Congress, Arizona, were "undeniably warm." Mrs. George Possauer delivered a ringing indictment of Miami when she described it as "the worst [camp] I have ever seen. The sun shines all day but the nights are colder than the ones in Alaska and the dust and dirt are nearly knee deep."[24]

In general, the miners complained about the inhospitable climate and ignored the beauty seen by men like Alers and Galloway. Bad weather hindered mining operations, causing hardships for the labor-

---

[21] Diaries of Major O. S. Alers, especially June, July, and August, 1901, and 1902, O. S. Alers Papers, WHRC.

[22] Richard A. Pomeroy to Will, Georgetown, Colorado, May 13, 1877, Stevens Mining Company Papers, DPL; R. Harvey to Mrs. R. and Miss Lucy Harvey, April 12, 1874, "Caribou," Spring Papers, WHRC.

[23] W. F. Staunton, "Memoirs," I, 96, William Field Staunton Papers, Special Collections Department, University of Arizona Library, Tucson (hereafter cited as UA); Parsons, *Journal*, January 27, 1881, p. 202.

[24] Staunton, "Memoirs," I, 96, Staunton Papers, UA; "Stay Away from Arizona," *Miners' Magazine* 17 (February, 1916), 6.

ing miners, whereas good weather passed unnoticed in the hustle and bustle of everyday life. Working eight- to twelve-hour shifts, miners spent little time contemplating nature's splendor.

In most of the mountain camps winter began as early as mid-October. On October 12, 1880, Eben Olcott, a Silverton, Colorado, mine manager, wrote that he was "still snow bound—It has only been storming for 98 hours and 20 minutes and still she has not given out." Five days later a temporary lull in the blizzard gave him an opportunity to complete winter preparations, specifically, to string about the camp a rope railing as both a handhold and a directional guide for the men.[25]

Seasonal shutdowns routinely occurred on mines that were in a developmental stage. Their operations failed to justify the increased costs of winter work, which usually necessitated covering the shaft or tunnel and the ore house, lest they fill with snow. In any event, winter work exposed the miners to suffering. In Nevada's White Pine District, winter closures were common because the prolonged severity of the weather and the topography made winter work unprofitable and hazardous. Consequently, miners who remained in the district during the winter had to secure temporary employment, which often was in short supply.[26]

In Colorado's high Rockies winter was everywhere severe. For example, in 1913 a record snowfall on central Colorado's eastern slope measured four and a half feet in Georgetown, reached five feet in nearby Idaho Springs, was so dense in Central City that lights came on at midday, and was reportedly eight feet deep atop Lookout Mountain. This sudden accumulation crushed some buildings and moved others from their foundations.[27]

Elsewhere the situation was the same; in 1875 a Virginia Citian noted that local citizens had been "snowed in nearly the entire winter." Such weather posed both physical and psychological dangers for the residents of the mining frontier. The physical dangers were ob-

[25] Eben E. Olcott to Pheme [Euphemia], October 12 and October 17, 1880, Eben E. Olcott Papers, WHRC.

[26] Ernest LeNeve Foster to Messrs. Baltzer and Lechtenstein, "Report on Mabel Mine," June 18, 1887, Ernest LeNeve Foster Papers, DPL; Jackson, *Treasure Hill*, p. 51.

[27] "Deepest Snow since the Glacial Period," newspaper clipping in Jesse Summers Randall Papers, SHSC.

vious: frostbite, exposure, and snow blindness. As early as 1880 Olcott explained to his sister that colored goggles were an "absolute
necessity" in Silverton. The psychological impact of the long, often
dreary, and isolated winter existence cannot be easily ascertained.
Nevada pioneer Rachael Haskell complained in her diary that "this
eternal covering of snow which continually meets the gaze so glitter-
ing must prove hurtful to an organ whose natural color should be
green to look upon." While Rachael found solace in her fire, her children, and her cheerful sitting room, miners had few such diversions
as they trooped off to the mines and spent yet another day underground.[28]

The most dramatic danger of the winter season was the snowslide. Along with fire, it was one of the great natural disasters that be-
fell mountain mining towns. Like fire, it swept through communities,
individual cabins, and mines with little or no forewarning; like fire, it
left only tragedy in its wake. Miners inadvertently increased the risks
of avalanches by denuding the mountainsides of trees, which they
used for surface construction, firewood, and mine timbering. Jesse S.
Randall, long-time editor of the Georgetown *Courier*, maintained a
clippings file that dealt specifically with snowslides occurring between
1874 and 1925. Often working away from town and up in the mountains, miners were especially vulnerable to slides. Writing in 1884,
one reporter concluded that these phenomena were the "most ruthless death dealers," in part because they struck so "unexpectedly."[29]
Lucky victims of slides received a few cuts and bruises, others suffered from exposure and frostbite, and the remainder died, usually
asphyxiated beneath a pile of snow, dragged down a mountainside, or
thrown over a precipice.

Miners found the change from spring to winter less dramatic
than in the east. Winter storms often lasted into April or May; in
1877, Richard Pomeroy recorded fully twenty feet of May snow and

[28] [Gideon] Anthony Hamilton to [S.] Frank [Dexter], February 6 [1875],
[Gideon] Anthony Hamilton Papers, HL; Eben E. Olcott to Pheme, October
17, 1880, Olcott Papers, WHRC; Richard G. Lillard, ed., "A Literate Woman
in the Mines: The Diary of Rachael Haskell," *Mississippi Valley Historical
Review* 31 (June, 1944), 84, 85, 94, entries for March 3, 4, and 26, 1867.

[29] Newspaper clippings on snowslides, Randall Papers, SHSC. See also
J. M. Goodwin, "Snowslides in the Rockies: A Perilous Study," *Overland
Monthly*, ser. 2, 29 (April, 1897), 384–385.

avalanches galore near Georgetown. The late storms posed serious problems because snowslides invariably accompanied them, and the build-up of wet, spring snow meant almost certain structural collapse if it was not shoveled from the roofs. Melting rapidly under the warm, spring sun, the snow quickly turned into slush, mud, and flood.[30]

As the weather moderated, spring and summer brought clear, warm days, the blooming of flowers, renewed communication with the outside world, and the chance to escape the strictures of winter's confinement.[31] With spring, summer, and early fall following in rapid succession, miners could count on three temperate months, coupled with good weather in parts of April, May, September, and October. Yet even then storms could and did raise havoc. Windstorms achieved lasting fame when William Wright labeled them "Washoe zephyrs." The effect of the zephyrs, wrote a half-jesting Wright, was to fill the air "with dust, rags, tin cans, empty packing-cases, old cooking stoves, all manner of second-hand furniture, crowbars, log chains, lamp-posts, and similar rubbish." Their results could be as innocuous as a few hats blown from the heads of the unsuspecting or as serious as a roof lost or a house blown down. Theodore Van Wagenen reported that he and a companion were amused by the activity of a mountain windstorm until it blew in the window of their room and deposited dirt and debris everywhere.[32]

Rain, hail, and sometimes even snow were part of the miner's summer lot in the mountains. July and August often brought daily showers, which one Coloradan noted usually began around noon, though sometimes as late as two o'clock, and continued for several hours. The only alternative to a good drenching was to find shelter quickly. In 1875, Alfred P. Camp recorded a mountain storm that began with driving snow and hail, which soon changed to rain. Forced to take shelter under a pine tree, he resumed his journey only to discover that the trail was so slippery in the steep places that he "was obliged to dismount and lead the horse." Meanwhile on Colorado's

[30] Richard A. Pomeroy to C. R. Pomeroy, May 13, 1877, Stevens Mining Company Papers, DPL.

[31] Harry C. Cornwall, "My First Year in the Gunnison Country," ed. Duane Vandenbusche, *Colorado Magazine* 46 (Summer, 1969), 234–235.

[32] William Wright, *The Big Bonanza*, p. 188; Theodore F. Van Wagenen, "Colorado," pp. 11–12, Theodore F. Van Wagenen Papers, DPL.

western slope, Albert B. Sanford recorded hailstones that were the size of hickory nuts and soaking afternoon rains that he described as a "usual" occurrence.[33]

While these summer storms caused considerable inconvenience and frustration, they also produced actual dangers. Often accompanied by lightning, they claimed occasional fatalities. Few men had the singular good fortune of Eben Olcott, who apologized to his sister for failing to write by noting: "I believe I did not write but last Sunday night I had a slight stroke of lightning in that magnificent thunderstorm we were in for several hours. It was not more than a good shock from an electric battery merely making my head tingle for a second or two."[34] However, the real danger from the spring and summer rainstorms came from mud and the notorious flash floods that swept through the arid and elevated countryside. Irritating and hazardous, mudslides were the spring and summer counterparts of the avalanches. At best, they provided an annoying break in the daily routine. For example, in 1903 a Georgetown slide cascaded down Griffith Mountain and spilled across the railroad tracks into town. Mrs. Schuler discovered that it had brought about a foot of sludge into her yard, leaving her the unpleasant task of cleaning up. In 1872 Elmer Faunce reported another that had muddied yards and torn down trees; sheer luck had kept it from oozing into a house owned by Robert Old. Brownsville, Colorado, was the scene of a famous mudslide, when slipping slag and waste from the mines swept through the dying community. With the local mines inactive, the residents had neglected to inspect the dumps and thus had lost the chance to avert the calamity.[35]

Mountain storms also produced devastating flash floods. An 1896 deluge killed twenty-nine persons, destroyed property throughout Jefferson County, Colorado, swept away bridges, and even tore up railroad track near Golden. Tombstone's Parsons reported personal

[33] Wm to Lucia, July 21, 1866, Wm[William] Letters, DPL; Journal of Camp, Camp Papers, pp. 22–23, SHSC; Diary of Albert B. Sanford, July 26 and August 9, 17, and 25, 1881, Albert Byron Sanford Papers, SHSC.

[34] "Killed by Lightning," *Denver Rocky Mountain News*, September 7, 1860; Eben E. Olcott to Pheme, June 26, 1880, Olcott Papers, WHRC.

[35] Newspaper clippings on snowslides, Randall Papers, SHSC; Elmer Faunce to Robert O. Old, Georgetown, May 31, 1872, William A. Hamill Papers, DPL; George Rowe, "The True Story of a Rich Silver Mine," p. 2, George Rowe Papers, DPL.

privation and isolation stemming from storms in August, 1881, and Frank Crampton chuckled about the consequences of a 1916 storm in Goodsprings, a camp along the Nevada-California border. The resultant flash flood made an island of the community's red-light district and captured both the fancy ladies and their unsuspecting patrons. He chortled that while no one perished, the captives were given a "warm" reception by their spouses and the community at large.[36] No doubt, some of these unfortunates might have preferred to have been swept away at flood tide.

By late August or early September, the residents of the mountain communities could expect an occasional snowstorm, signaling the advent of winter and another seasonal cycle. It is little wonder that the Great Basin and Rocky Mountain miners complained about the weather. Normally short of water, they sometimes endured torrential downpours; needing a temperate climate to press their extractions, they endured extremes of temperature. Often lowlanders, they struggled to adjust to the high altitudes of places like Virginia City, Jerome, Leadville, or Caribou. Combined with the ever-present mine dust, these conditions facilitated the spread of infectious diseases and contributed to the high mortality rate.

The discomforts and dangers miners faced, then, were legion. Both the diverse backgrounds of their neighbors and the unfamiliar climate and landscape confronted them with challenges and changes. How were they to live with this polyglot population at what often seemed the end of the world? They would work out an answer in the mining communities they formed. Miners cooperated in the construction of a transient mining-town society where they could establish their families and seek their livelihood.

[36] Newspaper clippings on floods, Randall Papers, SHSC; Parsons, *Journal*, August 25, 1881, p. 250; Crampton, *Deep Enough*, pp. 249–250.

2.

The Miners' Towns and Families

SUCCESSFUL lode mining required capital, careful management, and a force of trained miners. Working full time in the extraction process, the miner depended upon others to supply his basic needs—food, clothing, shelter, and recreation—and these needs spawned communities. The miners' towns, largely drab and mundane, nonetheless provided enough excitement and glitter to lend credence to the myth of the West. These cities, which experienced many of the same problems as other nineteenth-century urban communities, appeared to be isolated island communities scattered throughout the Rocky Mountain and Great Basin West. Linked in people's minds with the product of their mines, they attracted the attention of American society. Though remote, they were often visited; though isolated, they boasted a remarkably cosmopolitan population. The miner regarded his town as the center of his off-work universe.

The towns themselves were somewhat haphazardly located near mineral deposits. While early settlers attempted to establish order in the development of the townsites, they were rarely successful because control over or title to the land was difficult to obtain in the "rush." Miners' desire for proximity to the strike without unnecessary inaccessibility gave rise to helter-skelter communities, which typically followed the contours of the land. Newcomers erected tents, crude cabins, wagons, and early stores in canyons, along stream beds, and on the sides of sloping hills and mountains. Subsequent urban order invariably was constrained by a naturally determined pattern of growth.[1]

Although residents became inured to the distinctive qualities of their habitat, visitors noticed first the smells of the yet unseen settle-

[1] Eben E. Olcott to Pheme, Lake City, Colorado, May 30, 1880, Eben E. Olcott Papers, Western History Research Center, University of Wyoming, Laramie (hereafter cited as WHRC); Duane A. Smith, *Rocky Mountain Mining Camps: The Urban Frontier*, pp. 42–50.

ment. In the mountainous landscape, indifferent sanitation brought the pungent fumes of human and animal ordure, rotting garbage, and decaying organic material to the nostrils long before anyone actually saw the community. The aroma of burning wood and coal blended with the garlic-like fragrance emitted by the smelters. Even in the nineteenth century these smelter fumes were known to "more or less affect the health of all who resided in the town."[2]

What the nose first sensed, the eyes and ears soon beheld and heard. Perceptive observers noticed the discoloration of streams, attesting to the presence of stamp mills. Discarded machinery, abandoned prospect holes, dumps of active and inactive mines, billowing smoke, and the roadside litter of an urban, industrial society finally alerted even the inattentive. Then, rounding a curve or cresting a hill, visitors saw the town itself. Now its distinctive clamor was unmistakable. There were the dull booms of underground explosions, likened by some to the beating heart of the camp. Joining these were pounding stamp mills, groaning machinery, rumbling wagons, clattering hoofs, clanging and scraping hammers and saws, strains of music from dance halls, and chattering humanity engaged in myriad activities. The pandemonium was unending, since producing mines operated around the clock, and purveyors of pleasure kept pace with the mines. As Si Warman wrote of Creede, Colorado, "It's day all day in the daytime, there is no night in Creede."[3]

Students of the mining frontier have often viewed the mining camps in anthropomorphic terms. The camps seemed to have a distinct life-cycle, beginning with discovery, giving way to boomtimes, and then either withering away or entering a period of "maturity" ultimately followed by inevitable decline. The discovery of precious minerals usually precipitated an immediate rush and boom, whereas base metal discoveries rarely produced such results, in part because their exploitation required careful planning and extensive capitalization. Springing up spontaneously, the population of gold and silver camps soon exceeded the available supplies and services. Following

[2] "Among American Miners," *Chambers Journal*, ser. 5, 8 (1891), 572; "Eureka, Nevada," *Engineering and Mining Journal* 17 (1874), 213.

[3] "Interview with T. S. Ashley," *Gunnison News-Champion*, December 2, 1943, reprinted in Henry F. Lake, Jr., ed., "Remembrances of Pioneer Days in Gunnison County."

the prospectors and miners came a host of merchants, shopkeepers, craftsmen, liverymen, gamblers, saloonkeepers, innkeepers, and prostitutes. This hodgepodge of humanity remained in town until the mineral deposits were exhausted or until some other strike beckoned them to greener pastures.[4]

The history of Virginia City illustrates the rapidly changing conditions in these communities. Within a year after mining began in 1859, permanent frame buildings replaced makeshift structures. Eliot Lord reported that in the fall of 1860 the city boasted 202 firms, including general merchandise stores, tobacco shops, druggists, stationery shops, saloons, restaurants, boarding houses, a hotel, various smithies, cobblers and carpenters, watchmakers, physicians and a dentist, lawyers, assayers and a surveyor, livery stables, laundries, a bath house, a theater, a dance hall, a school, and a post office. The variety and extent of this list suggest that within a short time after its initial rush Virginia City had become an urban center. By the mid-1860's distinct occupational and socioeconomic classes had appeared. This social stratification was apparent in the neighborhoods reserved for residences of the merchant and mining elites, a distinct business district, Indian and Chinese ghettos, a red-light district, and residential suburbs for laborers. By then it had assumed the earmarks of a nineteenth-century city.[5]

No aberration, its development paralleled that of other mining communities, even the smaller ones. Frank Crampton recalled that early twentieth-century mining camps were

all cut from the same pattern, and each was near a group of mines. They had narrow streets with a few stores, eating places run by Chinese who knew how to cook, sometimes a bank, not often an "opera house," but always saloons and gambling places. A few homes were built on narrow streets away from the main drag; their number was small and often the owners took in boarders, mostly hard-rock stiffs who worked at near-by mines. There was always the "line" or "Stockade" with its dance halls, parlor houses, and cribs.[6]

These smaller towns that Crampton described experienced an initial

---

[4] Rodman W. Paul, *Mining Frontiers of the Far West, 1848–1880*, pp. 9–10; Smith, *Mining Camps*, pp. 42–50.

[5] Eliot Lord, *Comstock Mining and Miners*, pp. 72–73, 94, 198–199.

[6] Frank A. Crampton, *Deep Enough: A Working Stiff in the Western Mine Camps*, p. 9.

surge of growth, but lack of adequate ore reserves, inaccessibility, or overshadowing by a neighboring city stunted their growth.

During the mines' years of peak production, the cities became commercial and cultural *entrepôts*, praised for everything from their cuisine to their spectacular scenery. They displayed a diversity of human enterprise that stamped them as cosmopolitan by nineteenth-century American standards. All resembled the major American metropolitan areas more than they did comparably sized rural communities. Unlike other cities, however, they owed their prosperity and stability entirely to the uncertain extraction of minerals from the earth. Their boomer psychology fostered unbounded optimism, which projected present realities into a roseate future and ignored the finite nature of mineral deposits. The faith that there would be plenty more where that came from explained why these communities developed so rapidly. Expecting the very latest in material conveniences and social progress, their citizens obtained telegraph, telephone, and railroad connections as well as electrical, water, and sewage disposal systems, uncommon conveniences in places of comparable size.[7]

Gradually all mining towns declined unless they reduced their dependence upon ore removal. Where mineral deposits were extensive, deterioration occurred over several years or perhaps decades, as was the case in Leadville and Virginia City. Elsewhere, decay was more rapid. In declining towns, the boomtime necessities became hard-time frills. Public and private services degenerated as the money with which to pay for them disappeared.

Toward the end of the nineteenth century, copper mining attracted hard-rock miners. Copper miners found themselves in a slightly different social environment than other miners, because copper camps were often company towns. While company towns have had a long and varied history in the United States, the corporate paternalism that became their trademark appeared first in Lowell, Massachusetts. Thereafter, the company-town contrivance arose in single-industry communities around the nation. Two aspects of western coal and

[7] *Among the Silver Seams* (Georgetown, Colo.: *Georgetown Courier*, 1886), pp. 10–12, Mining File, WHRC; Smith, *Mining Camps*, pp. 99–123; Paul, *Mining Frontiers*, pp. 72, 104–105; Ray Colwell, "Cripple Creek: Some Recollections of My Boyhood There," pp. 1, 16, Alfred B. Colwell Papers, WHRC.

copper extraction encouraged the development of company towns around these industries rather than around other types of mines. First, the relatively large initial investment required to open such mines both enhanced the value of the stability company towns afforded and minimized proportionally the further expense their construction entailed. Second, and related to the high initial costs, coal and copper mines were developed only if the ore deposits were extensive enough to guarantee long-term operations. Therefore, the towns the companies built could also be expected to be long-term investments. In short, by constructing company towns, corporations could promote, protect, and control their mining investments. Moreover, the copper industry was promoting company towns at precisely the same time the union movement, especially the Western Federation of Miners (WFM), was making its most determined effort to organize hard-rock miners. This coincidence in timing was hardly surprising, since the company town gave management additional leverage in dealing with both labor organizers and their own employees.

The one study addressing the phenomenon of company towns is avowedly revisionist. In challenging the traditional assumptions about the nature of company towns, its author, James B. Allen, stresses the physical benefits provided and minimizes the intangible impact of corporate paternalism. Significantly, however, the era of corporate concern for the welfare and physical well-being of employees documented by Allen postdates the era under study here.[8]

Early twentieth-century muckrakers and labor sympathizers scrutinized and often criticized company towns. For example, a reporter for the socialist *New York Call* described Bingham, Utah, as a "modern inferno, a veritable slaughter pen where the workers are slaughtered and maimed wholesale." The *Call* denounced Utah Copper for an inadequate wage scale and for violations of Utah's eight-hour-day law. It concluded its description of the camp and the corporate policies with a graphic account of typical miners' quarters: "Dark, windowless, floorless, they are constructed out of powder boxes, the sheet metal of tin cans and such odds and ends of timber as the miserable builders are able to pick up without getting arrested." The *Miners' Magazine*, official organ of the WFM, periodically re-

[8] James B. Allen, *The Company Town in the American West.*

ported on conditions in company towns. For instance, it quoted a charge by the *Mining and Scientific Press* that it was a "blot on our democracy that citizens should live among such surroundings" as were found in Jerome, Arizona. The *Miners' Magazine* went on to contrast the despicable conditions in Jerome with the luxury and opulence that its copper permitted the mineowners.[9]

Even the casual observer discerned important differences in the physical appearance of company towns compared with non-company communities. Under company supervision, houses were built in systematic patterns, painted alike, and located along streets leading inexorably to the mines, mills, and smelters. These monotonous dwellings were nonetheless better than those available in independent mining towns, because the companies constructed durable houses that might even include indoor plumbing, a luxury elsewhere.[10]

Concerned with community morality and worker productivity, corporations regulated saloons and restricted or, preferably, eliminated gambling halls and red-light districts. Local businesses operated at the companies' pleasure, and corporate officials licensed them, to ensure a limited number of accountable establishments. Ordinarily owning and operating boarding houses and hotels, the companies also ran all-purpose stores that supposedly offered goods at reduced prices. When coupled with corporate ownership of residential lands, these stores created closed economies in which employees extracted minerals and in return received the goods and services necessary for survival.[11]

In this atmosphere, resident management exerted subtle control over the social, moral, cultural, and political affairs of the communities. The companies built recreational facilities that housed bowling alleys, pool and card tables, libraries, gymnasiums, and clubs and provided supervised leisure-time activities. The libraries did not sub-

[9] "Dire Exploitation of Copper Trust's Slaves Unites 24 Nationalities in Bond of Solidarity," *New York Call*, October 6, 1912, Strike Scrapbook 2, Daniel Cowan Jackling Papers, Bancroft Library, University of California, Berkeley (hereafter cited as BL); "The Clarks' United Verde Interests," *Miners' Magazine* 19 (June, 1918), 5.

[10] Russell R. Elliott, *Nevada's Twentieth-Century Mining Boom: Tonopah, Goldfield, Ely*, pp. 216–222; J. C. Kinnear, "Fifty Years in Mining, 1907–1957," pp. 27–28, J. C. Kinnear Papers, WHRC.

[11] Elliott, *Nevada's Boom*, pp. 221–222.

scribe to "subversive" literature from unions, social reformers, or radicals. Describing Bisbee, Arizona, in 1894, Eben Olcott noted that everyone was a company employee, including the minister, who doubled as postmaster. He characterized Ben Williams, the Phelps Dodge superintendent, as a "splendid man, most interested in everybody and everything," not an entirely reassuring portrait.[12]

Local politics did not exist in company towns; instead, all local policies were determined by corporate fiat. Consequently, there was greater interest in national, state, and county governmental affairs, which provided alternative arenas for involvement. While company towns could offer comparatively better systems of streets, sewage, public services, and residential housing, they sacrificed participatory democracy for corporate paternalism. Political and individual freedoms were replaced by a benign dictatorship that operated independently of local communities and their interests. Significantly, because of their autocratic nature, these towns did not facilitate the acculturation of immigrants, Mexican-Americans, and Indians who worked in the copper camps. The distance between concerned paternalism and oppression rested on the whims of absentee owners, whose primary consideration was the business ledger.[13]

Requiring strenuous and monotonous labor, mining was judged an occupation suited only to men. Women and children were a minority in the booming towns, an exception in remote camps. Consequently, bachelors, actual and circumstantial, shaped the general temper of the towns. Each June the youthful and masculine character of life was reinforced by the arrival of college students drawn by the lure of mining and its high wages. A summer's work in the mines brought enough money to finance another year's education, and for aspiring engineers and assayers it provided excellent on-the-job experience. While single girls rejoiced at the influx of beaus, other residents took less comfort in their arrival. Intoxicated with the myth of the West, the collegians not only provided unwanted competition for the few marriageable women, but also often engaged in the self-assertive pranksterism of the collegiate way. However, the miners were well-

---

[12] Eben E. Olcott to Euphemia, February 24, 1894, Olcott Papers, WHRC.
[13] Elliott, *Nevada's Boom*, pp. 221–222.

known practitioners of their own forms of ribald and mischievous shenanigans and, no doubt, gave better than they got.[14]

Undergoing the same pattern of development as the towns, miners' accommodations were initially crude and improvised. Canvas tents became their first homes in most camps, but rocks, trees, crevices, caves, and even "buffalo-chip" adobes sufficed for the desperate early arrivals. As populations grew, enterprising souls appropriated litter for construction purposes; in the early twentieth century, Goldfield and Tonopah rushers fashioned dwellings from bottles, barrels, and oil cans. Fortunate individuals arrived with wagons, which provided temporary habitations as well as a source of lumber for subsequent home improvements. Providing only a modicum of protection from hostile elements, snakes, and wild animals, these abodes were soon abandoned in favor of conventional housing.[15]

Living quarters remained scarce but showed marked improvement in quality as the camps entered the boomtown phase. Reaching Leadville in the late 1870's, Henry Wood searched in vain for lodging, until finally a sympathetic shopkeeper permitted him to sleep on the counter. Thirty years later Frank Crampton and his companions spent an entire day trying to find a place to stay in Cripple Creek. Rejecting bunkhouses, they located a boardinghouse that had a vacant room and three cots. The housing shortage changed drastically once a community entered the years of decline, however. Property owners were among the first to suffer from reversals in mining; prices fell and rental properties became impossible to lease.[16]

[14] Interviews of the Colorado Pioneers (John Griffith and Dominick Faussone), State Historical Society of Colorado, Denver (hereafter cited as SHSC); Journal of Alfred P. Camp, July 27, 1875, Alfred P. Camp Papers, SHSC; Thomas D. Sanders and Charles C. Stemmer, "Hardships and Perils on the Western Frontier or Fifty Years of Pioneering," p. 87; R. Colwell, "Cripple Creek," pp. 21–22, Colwell Papers, WHRC.

[15] Wolfe Londoner, "Western Experiences and Colorado Mining Camps," *Colorado Magazine* 6 (March, 1929), 69; Henry C. Morris, *Desert Gold and Total Prospecting*, p. 27; James K. Hastings, "A Boy's Eye View of the Old Southwest," *New Mexico Historical Review* 26 (October, 1951), 293.

[16] Henry E. Wood, "I Remember," p. 30, Henry Ellsworth Wood Papers, Huntington Library, San Marino, California (hereafter cited as HL); Crampton, *Deep Enough*, p. 41; Ernest LeNeve Foster to John M. Phillips, Georgetown, January 26 and April 9, 1886, Ernest LeNeve Foster Papers, Western History Department, Denver Public Library (hereafter cited as DPL).

Customarily the bachelor miners lived in rooming houses or dormitory-like bunkhouses. In 1880 Virginia City's standard room measured ten feet by twelve feet and rented for about thirty dollars per month, board included. At approximately a dollar a day, these prices were nearly a third lower than in the bonanza days of the 1870's. Individual stoves heated the rooms, and doors and windows provided rudimentary ventilation. In Independence, Colorado, a Cripple Creek suburb, one miner remembered paying ten dollars a month for a mere cubicle. "It contained a bed, an affair on legs which had two small drawers and which supported a wash bowl and pitcher, and a stove. . . . The out wall of the room was made of two thicknesses of boards with tarred paper in between. . . . The inner walls would not have held a nail. They were vibrant tremulous things which acted as sounding boards and recorded and amplified whatever went on in the house." However, in nearby Victor, cots alone had once rented at a dollar per day and had been kept in continuous use by the changing shifts. For single men or mining-camp "widowers," rooming-house living provided the essentials: shelter and a bed.[17]

Other single men preferred to live in cabins or crude houses, which they built or rented. George Parsons of Tombstone erected ten-foot by twelve-foot shacks that he leased to miners at ten to fifteen dollars per month. Costing Parsons less than one hundred dollars apiece, these cabins returned a handsome profit in their first year. Parsons himself shared what he termed a "rough house," which was "simply roof and sides with openings all over through which the wind came freely." One Cripple Creeker described his dwelling as having one thickness of boards covered with corrugated iron, "an oven in summer, an ice box in winter." At Hillerton, Colorado, Albert Sanford and his father purchased a "neat log cabin" that had been a newspaper office. It housed four miners and boasted window glass, shingles, and a wooden floor. These sparsely furnished shacks contained old crates for stools and tables, a stove or fireplace, and shelving for personal belongings and foodstuffs. Providing only a minimum of shelter,

---

[17] Lord, *Comstock Miners*, pp. 372–373; Charles H. Shinn, *The Story of the Mine, as Illustrated by the Great Comstock Lode of Nevada*, pp. 253–254; Leo J. Keena, "Cripple Creek in 1900," pp. 1, 6–7, Leo J. Keena Papers, SHSC; S. E. Poet, "Victor, Colorado—The City of Mines," *Colorado Magazine* 10 (May, 1933), 106–114.

they nonetheless offered a sense of privacy unavailable in the flop-houses.[18]

Married men had little choice except to build or rent a simple one- or two-room cabin. One critical observer described these un-painted dwellings as "decidedly inferior externally" to similar homes he had known in the East. The wives added touches of domesticity with curtains at the windows, rugs, painted or wallpapered interiors, and knickknacks. James Galloway, the Comstock miner, bought a lot for $225, upon which he built a house costing $800. Completed within ten days, it was basic in design and was probably constructed of ver-tical beams held together with battens.[19]

In 1907 one young Goldfield miner and his wife, after a long search, finally rented for forty-five dollars a month a one-room frame house with an attached lean-to that served as the kitchen. The house lacked indoor plumbing and had the customary outhouse. Its most ad-vantageous feature was its location, not far from the mines. In Cripple Creek one Midwestern farm family (the O'Briens) found the housing shortage a new and frustrating experience. In her desperation Mrs. O'Brien camped on the steps of a house that was to be vacated. After taking possession as its former occupants left, she managed to trans-form the four-room abode into a warm and pleasant home. Another Cripple Creek family dwelt in a "good four-room frame house, with peeled slabs on the outside so that it looked like a log cabin, but wasn't." Such families invested considerable sums of money in their homes and, doubtless, were better sheltered than most.[20]

Though scarce, women were important on the mining frontier; every major community had its "line" or "stockade" where "working stiffs, and businessmen of the town, too, went for relaxation and com-panionship of the girls who lived there." Prostitutes' lives and life-

---

[18] George Whitwell Parsons, *The Private Journal of George Whitwell Parsons*, February 17, 1880, p. 92; R. Colwell, "Cripple Creek," pp. 18–19, Colwell Papers, WHRC; Ross B. Moudy, "The Story of a Cripple Creek Miner," *Independent* 57 (August 18, 1904), 380–382; Diary of Albert B. Sanford, June 14, 1881, Albert Byron Sanford Papers, SHSC.

[19] Lord, *Comstock Miners*, pp. 372–373; John Debo Galloway, *Early Engineering Works Contributory to the Comstock*, pp. 16–17.

[20] Kinnear, "Mining," p. 9, Kinnear Papers, WHRC; Pearl O'Brien Col-well, "Pearl," p. 12, and R. Colwell, "Cripple Creek," p. 4, Colwell Papers, WHRC.

styles, however, received scant attention except as legal and moral issues. Violence, vagrancy, and occasional tragedies in the red-light district gave local editors an opportunity to lecture their communities on immorality and to reiterate maxims of proper behavior, but such tragedies rarely encouraged thorough analysis of the women's situation.[21]

Fortunately for later historians, some observers described these rows of cabins "with gaudily-furnished rooms." Most cribs were little more than shacks, furnished with a bed and perhaps a chair; a few "courtesans" lived in elaborate bordellos like Cripple Creek's Homestead. Little was known of the women's origins. Some came west with dance troupes or theatrical companies; others found themselves deserted or widowed in a hostile environment; many of the Orientals were virtually slaves; a few women were drawn into their state by callous and greedy friends or relatives; and still others came west to escape poverty or to recoup lost fortunes.[22]

Living in anonymity, the harlots found solace in alcohol, opium, and morphine, all openly available in most substantial mining communities. Venereal diseases were ever-present dangers for these women, as were depression and desperation leading to suicide. Some of the women found husbands among the citizenry and settled into respectable lives. A very few may have made enough money to return home and resume normal lives. As pictured by early photographers, the *filles de joie* were rough and rugged women; few could have been described as comely. As Frank Crampton recalled, they were "real women" who "did something wonderful to lonely men, and it was not the momentary pleasure in a bed. I was one of them, and I know." Their value so impressed him that he encouraged their residence in mining camps that he later operated as an engineer and consultant.[23]

As societies matured, matrons and reformers campaigned to

[21] Crampton, *Deep Enough*, pp. 9, 50.

[22] Parsons, *Journal*, July 16, 1880, p. 139; Lord, *Comstock Miners*, pp. 198–199; Keena, "Cripple Creek," p. 1, Keena Papers, SHSC; M. A. Deering, "My Pioneer Experiences in the West," *Gunnison News-Champion*, May 2, 1935, reprinted in Lake, "Remembrances"; Frank S. Hastings, *A Ranchman's Recollections: An Autobiography*, p. 101.

[23] Crampton, *Deep Enough*, pp. 50, 62–64, 68–70, 139–143, 162–163, 192–194; Fred M. Mazzulla, *Brass Checks and Red Lights*.

eradicate prostitution, or, failing that, to penalize lewdness, open solicitation, and vagrancy. Total elimination of the practice proved nearly impossible because bachelors and tenderloin investors thwarted the forces of Victorian decency. In periods of urban decline, prostitution suffered the same reversals as other segments of the mine-linked economy.[24]

While wives, daughters, and sisters soon followed their men to the mining districts, they remained few in number and consequently acquired rights and privileges elsewhere denied them. For example, both employment and divorce lost their stigmas on the frontier. There were even scattered reports of women working in the drifts; one such, Mrs. D. H. Elder of Victor, claimed to have mined for two years. In Arizona, Frances Douglas (DeKalb) actually located claims and supervised assessment work done on her Roadside Mine.[25]

Ordinarily coming west to join loved ones, women accepted the rough life in order to establish or reunite the family. While commentators extolled the excitement and wonder of life in the mining regions, it was commonly agreed that life there was hard for women. In addition to the usual problems of adjustment, they faced housing shortages, high prices, and assorted discomforts. In remote areas the conditions were primitive. From Rico, Colorado, deep in the San Juans, the executor of a miner's estate tried to dissuade the miner's widow from coming west: "I never would advise any lady to come here unless it was absolutely necessary. We are on the extreme frontier with little or no conveniences, living very expensive, and our location very inaccessible [sic]."[26]

---

[24] Elliott, *Nevada's Boom*, pp. 66–68; Julian Street, "Colorado Springs and Cripple Creek," *Collier's* 54 (November 21, 1914), 16–17, 30; Bennett Cerf, "Trade Winds," *Saturday Review of Literature*, July 31, 1948, clipping in Robert Livermore Papers, WHRC.

[25] Mr. and Mrs. D. H. Elder to W. A. Morgan of WFM Local #80, May 28, 1903, *Official Proceedings of the Eleventh Annual Convention of the Western Federation of Miners of America, Denver, Colorado, May 25–June 10, 1903* (Denver: Western Newspaper Union, 1903), p. 144; Roadside Mine Papers, Frances Douglas Papers, Special Collections Department, University of Arizona Library, Tucson (hereafter cited as UA).

[26] E. D. Gardner Diaries, 1907–1910, especially April 19, May 1, and June 8 and June 9, 1907, E. D. Gardner Papers, WHRC; F. W. Raymond to Mrs. Mary Galbraith, September 17, 1880, George H. and Mary Galbraith Papers, DPL.

Operating on tight budgets, miners' wives often worked part time as seamstresses or laundresses; others operated boarding houses to supplement family incomes. During periods of labor unrest, they worked in hotels and laundries to augment meager savings and strike-fund allotments. Furthermore, mine accidents or job-related disabilities forced spouses into these occupations in order to support their families. They led lonely lives, as the mine shifts were long, and their men returned physically and mentally exhausted. The daily drudgery and isolation of this life moved Elizabeth Sayre to write, "I am deathly homesick and lonely," and later to say, ". . . it is forlorn as anything one can imagine for me here in Aspen." Women might even have to endure being sent back East to bear children or to cut costs while their husbands sought jobs in undeveloped camps; they felt forsaken and utterly frustrated.[27]

Women's daily worries ranged from financial insecurity due to declining productivity in the mines to the unsettling specter of widowhood. Each day they knew that their husbands might not return. Since few men could obtain high-priced life insurance, once widowed, women had no choice but to begin working wherever and whenever they could. As one Kokomo, Colorado, woman explained to the Gold Hill, Nevada, miners' union: "I am the widow of the late Sumner Whitney who was killed near Kokomo on Aug. 12. I have been left in almost destitute circumstances and I am now trying to save an encumbered home. To do so, I am compelled to raffle the family diamonds in order to raise money to lift the mortgage."[28] Corporate insurance and liability records indicate that others found themselves in similar straits. Ordinarily, circumstances put them at the mercy of the company's settlement offer, and before the appearance of liability insurance, widows had no recourse except through the courts.[29]

The most exasperating challenges for women involved trying to

[27] Interview with Albert J. Frantz, Denver, April 6, 1972; Diary of Elizabeth D. Sayre, October 29 and October 30, 1889, Hal Sayre Papers, Western History Collection, University of Colorado Libraries, Boulder (hereafter cited as CU).

[28] Mrs. M. A. Whitney to Secy. Miners' Union #47, Gold Hill, Nevada, n.d., Gold Hill Miners' Union Papers, BL.

[29] Kenneth L. Fahnestock to Thomas F. Daly, Leadville, July 5, 1898, John F. Campion Papers, CU; Mr. Dickey to Mr. Irving Bush, Central City, February 6 and February 26, 1898, Gregory Consolidated Mines Papers, CU.

raise their children and hold families together. In the industrial-urban society of the mining frontier, there were few provisions for children. Where even adult miners faced a seemingly inexhaustible list of hazards, every street, stream, tool, and machine posed potential danger for children. C. H. Hanington's reminiscences of his childhood in Central City give some indication of the escapades open to children. His playground ranged over the hills and into the mines strung out along Clear Creek Canyon. Roaming through old mine shafts was a favorite sport, as were expeditions into the Bobtail tunnel. In active mines like the Bobtail the excitement was heightened by the need to dodge the "numerous mule-drawn ore trains." Railroad trestles provided an equally dangerous alternative for the adventurous youths. Here was the nineteenth-century equivalent of the playground; opportunities to climb, crawl, cling, cajole, and dare abounded. Such ventures must have tried the patience of the most tolerant mothers.[30]

In this adult-oriented environment, boys quickly acquired the vices of gambling, cursing, drinking, and whoring. According to Ray Colwell, who grew up in Cripple Creek, they also formed juvenile gangs and battled over disputed territory. Organized by neighborhoods, they preyed upon unsuspecting adults and unorganized peers. Virginia City's John T. Waldorf and his "Savage gang" (named after the nearby mine) delighted in harassing the defenseless and hated Chinese and Indians.[31]

Young women posed distinctive problems for their concerned parents. With prospective suitors far outnumbering eligible women, most attractive and pleasant girls enjoyed active social lives. However, not all were so fortunate; the daughters of immigrant miners encountered both subtle and overt discrimination against "foreigners." Most girls became miners' wives, but some rejected the burdens and instability of that life and turned instead to harlotry. One young Cripple Creek girl followed this path after witnessing her mother's death and her father's daily absences.[32]

---

[30] C. H. Hanington, "Early Days of Central City," *Colorado Magazine* 19 (January, 1942), 3, 5, 13–14.

[31] R. Colwell, "Cripple Creek," pp. 21–22, Colwell Papers, WHRC; John Taylor Waldorf, *A Kid on the Comstock: Reminiscences of a Virginia City Childhood*, ed. Dolores Bryant Waldorf, pp. 115–116.

[32] R. Colwell, "Cripple Creek," pp. 21–22, Colwell Papers, WHRC;

The instability and mobility of life disrupted children's lives. One youngster later remembered that her family's move from Iowa to Cripple Creek had left her isolated and lonely. "I (the most timid of children) going to what seemed a tremendously large school, having no one, entirely on my own, no Iowa social standing of my grand-parents—whom I have worshipped—nor of my parents who could help me now—no one to explain what loneliness means without this experience." In this big and inexplicable country, children seemed insignificant; they did not control their environment and had no clear place in the future. Except for the attention given public schools, few special provisions were made for children in the bustle of the booming towns.[33]

As early as 1860 the Colorado camps of Nevada City, Missouri City, and Mountain City reported that they had employed teachers and had schools under construction. In communities ranging in size from Virginia City and Cripple Creek to Telluride and Brownsville (Colorado), citizens taxed themselves to provide rudimentary education for the young. Often the edifice was more important than the training received. Plagued by the frequent turnover and incompetence of instructors, the schools offered child care, a modicum of protection, and the rudiments of literacy. On this frontier, children remained an anomaly, even while elsewhere society glorified the future and youth.[34]

Thus, in this predominantly masculine and cosmopolitan setting, women and children had to accept, however grudgingly, the privations of frontier life and isolation. Residents of these remote mountain communities sought to re-create the social structure and public services they had left. For this reason, their towns often boasted the latest technological innovations in transportation and communication, as well as such typically American institutions as the public school.

Judge Albert S. Frost to Frank M. Ish, Cripple Creek, July 20, 1901, Albert S. Frost Papers, SHSC.

[33] P. Colwell, "Pearl," pp. 10, 12, Colwell Papers, WHRC.

[34] "From the Gregory Mines," *Denver Rocky Mountain News*, August 29, 1860; Lord, *Comstock Miners*, pp. 206–207; Elmer Faunce to Mr. Old, Georgetown, June 10, 1872, William A. Hamill Papers, DPL; R. Colwell, "Cripple Creek," pp. 6–7, and P. Colwell, "Pearl," p. 12, Colwell Papers, WHRC.

The uncertainty of mining, however, meant that without warning the entire community structure could dissolve as the people raced off in pursuit of a new bonanza. In these towns, their temporary world, miners encountered many dangers and developed compensating leisure-time diversions.

3.

The Perils and Pleasures
of the Miners' World

PROVIDING the backdrop for the human struggle, the camps, towns, and cities of the mining frontier contributed to the hazards of life and influenced the diversions of mineworkers and their families. Fire, disease, crime, unemployment, and inadequate public services complicated their lives, but for the most part neither the difficulties nor their proffered remedies were unique to the mining settlements. As members of communities, residents of even the newly formed camps had more in common with other urbanites than with isolated farmers or ranchers. Both the perils and the pleasures of the laboring miners developed within or were shaped by these communities.[1]

Only the exhaustion of all ore reserves (mines in "*borrasca*" instead of "bonanza") rivaled fire as the ultimate crisis for a community. Fire has always been a calamity, but on the mining frontier the arid environment, the slipshod buildings, and a slight breeze could turn isolated sparks into irrepressible conflagrations. Almost every substantial mining town suffered the depredations of fire.

Fires often inflicted severe hardships on the mine laborers. Since boarding houses were centrally located, they were likely to be touched by any fire. The married men rented or owned small dwellings constructed of highly combustible mine timbering, clapboards, and common paper. Furthermore, the miners' inferior socioeconomic status meant that their homes were sacrificed, while every effort was made to save the mines, the local businesses, and the homes of the merchant

---

[1] Charles M. Glaab and A. Theodore Brown discuss the problems of American urban living in their *A History of Urban America*. Gene M. Gressley provides an informative assessment of the literature on urban life in the Trans-Mississippi West in *West by East: The American West in the Gilded Age*, pp. 25–31, 45–48. Duane A. Smith's *Rocky Mountain Mining Camps: The Urban Frontier* is an excellent general discussion of urban problems on the mining frontier.

and mining elite. Even when their own homes were not destroyed, their lack of insurance and liquid assets meant that miners suffered privations if fire occasioned even a temporary shutdown of the mines.[2]

In this tinderbox environment, in which everyone and everything constituted a potential fire hazard, natural phenomena like lightning and spontaneous combustion posed special threats to communities. More often, however, fires stemmed from human carelessness and originated within the communities themselves. Livery stables were notorious firetraps, as were the assay offices, where careless construction of a furnace or chimney or inadequate precautions, given the constant use of the furnace at very high temperatures, could easily prove disastrous. Saloons too constituted persistent threats to the public safety. Crowded with milling and partially inebriated customers, they were shoddily constructed and contained storerooms filled with literal "firewater" that compounded the problems of controlling blazes. In addition, restaurants, hotels, opera houses, and even churches were subject to fires.[3]

Miners were at least partially responsible for many of the numerous fires that started in their boarding houses. These clapboard, sheet metal, and cloth structures could explode into flames; a smoldering cigarette, an obstructed or defective flue, an overturned coal-oil lamp, or, later, a defective electrical connection could initiate a raging fire. The notorious Virginia City fire of 1875 allegedly started with an overturned kerosene lamp in such a place, and one of Tonopah's early fires began in a rooming house.[4]

Finally, the mines, especially the surface workings, sparked or more frequently fueled blazes. Steam-powered hoists and pumps, potbellied stoves, lanterns, candles, tobacco products, and matches proved especially dangerous. One Comstock miner's son recalled that he

[2] Mabel Barbee Lee, *Cripple Creek Days*, pp. 87–101; [Gideon] Anthony Hamilton to Frank [Dexter], Virginia City, November 10, 1875, [Gideon] Anthony Hamilton Papers, Huntington Library, San Marino, California (hereafter cited as HL).

[3] Edward Robbins Howe to Katie, Stockton, Utah, July 1, 1873, Edward Robbins Howe Papers, Bancroft Library, University of California, Berkeley (hereafter cited as BL).

[4] Eliot Lord, *Comstock Mining and Miners*, pp. 325–328; John Debo Galloway, *Early Engineering Works Contributory to the Comstock*, pp. 17–18; Mrs. Hugh Brown, *Lady in Boomtown: Miners and Manners on the Nevada Frontier*, p. 82.

nearly triggered a fire in the Consolidated-Virginia change room when he inadvertently overturned a Franklin stove. His father and friends quickly reached the scene and extinguished the fire. Because flaming workings threatened both the community and its livelihood, battle lines always formed about the mines. This priority explains the similarity between the strategy of Virginia City's fire fighters in 1875 and their counterparts in Delamar, Nevada, in 1900.[5]

Except in company towns, local officials and the citizenry often dealt haphazardly with the problem of fire protection until disaster struck. Cripple Creek's experiences exemplified the danger of such negligence. On April 25, 1896, the *Cripple Creek Times* lauded the city's fire-fighting paraphernalia and quoted Mayor George Pierce, who had assured the residents that they could expect to "be free of this terror." When fire began that very day, the struggle quickly exhausted the single reservoir and left the community totally unprepared to confront the next conflagration on April 29. Chastened by these tragedies, Cripple Creek established and supported munificently an efficient and respected fire department.[6]

Raging fires left indelible impressions on those who beheld them. Mabel Barbee Lee survived the Cripple Creek fires and later recalled the terror that gripped her when the second alarm sounded: "I waited, almost paralyzed with dread. There could be no mistaking that direful sound. Again and again the desperate alarm was repeated until I could hardly hear the shots [the fire signal] for the pounding of my heart." In Virginia City the great fire of 1875 destroyed two thousand buildings, inflicted an estimated ten million dollars in property damages, and left hundreds homeless and destitute, but young John T. Waldorf remembered only the excitement of the occasion. "I can still shut my eyes and see the clouds of smoke and the blinding sheets of flame. In

---

[5] John Taylor Waldorf, *A Kid on the Comstock: Reminiscences of a Virginia City Childhood*, ed. Dolores Bryant Waldorf, pp. 43–45, 92–93; Anthony Hamilton to Frank, Virginia City, November 10, 1875, Hamilton Papers, HL; Lord, *Comstock Miners*, pp. 325–328; John M. Townley, "The Delamar Boom: Development of a Small One-Company Mining District in the Great Basin," *Nevada Historical Society Quarterly* 15 (Spring, 1972), 16–17.

[6] Lee, *Cripple Creek Days*, pp. 85–88; Ray Colwell, "Cripple Creek: Some Recollections of My Boyhood There," pp. 11–12, Alfred B. Colwell Papers, Western History Research Center, University of Wyoming, Laramie (hereafter cited as WHRC).

fancy I can follow the fire as it raced along the uptown streets. It was the greatest day of my five years of life. . . . It was better than a circus."[7]

Lawlessness, another common urban problem, resembled that found elsewhere. There were the juvenile gangs and town drunks, usually nuisances but seldom dangerous. Sometimes, however, they passed beyond being merely bothersome, as when drunks inadvertently started fires, engaged in brawls, discharged firearms, assaulted others, or got on a horse or in a buggy. Then they became more than public nuisances and were accorded little mercy. Public drunkenness led to the regulation and licensing of local saloons. Costing between twenty-five and fifty dollars, licenses of three to six months' duration forced saloon keepers to exercise a rudimentary supervision of their customers. Mine managers like John Campion occasionally assisted in limiting the effects of alcohol on the community by endorsing temperance, either as a matter of principle or as a way to increase worker productivity.[8]

Though the West's vigilante tradition has attracted much attention, its sway was usually brief and associated with the early boom days. Insofar as it affected laboring miners, it was linked with the labor-management confrontations of the late nineteenth and early twentieth centuries. Generally, the era of lawless conflict occurred before the laboring miners arrived. At most, these men served as occasional rank-and-file members of short-lived vigilante movements. In contrast to the vigilante image, miners and their families established a tradition of lawful conduct in their communities.[9]

[7] Lee, *Cripple Creek Days*, pp. 94–95; Waldorf, *Kid on Comstock*, p. 91.

[8] Lynn I. Perrigo, "Law and Order in Early Colorado Mining Camps," *Mississippi Valley Historical Review* 28 (June, 1941), 41–62; Sacrament Mining District Records, Licenses, vol. 19, Records of Colorado Mining Districts, State Historical Society of Colorado, Denver (hereafter cited as SHSC); Kenneth L. Fahnestock to John F. Campion, Leadville, July 24, 1902, John F. Campion Papers, Western History Collection, University of Colorado Libraries, Boulder (hereafter cited as CU). Alcoholism accounted for fully 9 percent of all diseases treated in the Storey County Hospital, Virginia City, Nevada (Lord, *Comstock Miners*, Table VII, p. 437).

[9] For a general discussion of vigilantism, see Richard Maxwell Brown, "The American Vigilante Tradition," in Hugh Davis Graham and Ted Robert Gurr, eds., *Violence in America: Historical and Comparative Perspectives*, pp. 154–226. Reference to vigilante activity in mining towns appears in Henry H.

Though transient, miners showed respect for orderly communities; though hungering for quick wealth, they honored the sanctity of private property. For example, residents of Colorado's early mining camps established law-enforcement offices to deal quickly with illicit activities and dubious economic practices. They strove to re-establish some semblance of law and order as they had known it back east. They selected their own local officials and created miners' courts that adjudicated civil and criminal cases. The disturbances familiar from Western lore were so infrequent that Leo J. Keena considered the commotion attending the accidental discharge of a gun in the Victor Club worthy of special mention. Nothing better illustrated the law-abiding character of the mining region than the Colorado municipal elections of 1867, when blacks first exercised the franchise "without the slightest molestation." This is not meant to suggest that violence did not occur, but rather to place it in perspective. Occasionally miners-turned-lawmen were killed in the line of duty, as was Central City's Dick Williams. Labor disputes or personal disagreements could provoke violence, as when W. F. Staunton's foreman killed a Congress, Arizona, miner, who, when refused work, had threatened the foreman. Violence was not, however, the norm.[10]

Much more threatening to miners than any violence were common urban diseases and pests. In fact, except for occupational accidents, illness and pests posed the most serious hazards miners and their families faced. Lacking adequate water supplies, sewage systems, and public health services, mining towns suffered infection and infestation. Human and animal offal provided both a breeding ground for germs and insects and food for the rodents. In the desert regions of New Mexico, Arizona, Utah, and Nevada, miners encountered poisonous reptiles, spiders, and scorpions. George Parsons concluded that, despite his efforts at eradication, "the creeping things are getting

Mason to Sister, Reno, August 14, 1874, Henry H. Mason Papers, BL; and Eben E. Olcott to Pheme, Silverton, September 19, 1880, Eben E. Olcott Papers, WHRC.

[10] Leo J. Keena, "Cripple Creek in 1900," pp. 5–6, Leo J. Keena Papers, SHSC; Frank Hall to Mother, Central City, April 14, 1867, Frank Hall Papers, Western History Department, Denver Public Library (hereafter cited as DPL); Arthur Cecil Todd, *The Cornish Miner in America*, pp. 157–158; W. F. Staunton, "Memoirs," I, 90–91, William Field Staunton Papers, Special Collections Department, University of Arizona Library, Tucson (hereafter cited as UA).

too numerous for comfort." Pestered by voracious flies, he noted sarcastically: "It's a curiosity to gaze at the beans at dinner. They're black."[11]

Men learned to shake their boots before putting them on to dislodge unwanted visitors, and they killed both snakes and scorpions whenever they encountered them. Rodents were a nuisance, as well as being carriers of contagious diseases, though in time cats ended this menace. On the Comstock cats drove the rats from the surface into the underground workings, where they thrived and multiplied. In the mines the rodents became almost pets. They ate the accumulated organic matter and also served as companions to the men; they gathered daily when the workers opened their dinner pails and shared in the scraps.[12]

In time, the cats replaced the rats as the number-one community nuisance. Six months after priding himself on the capture of a fine feline, George W. Parsons was damning the presence of screeching cats. He was so angered that on one occasion he drew his gun and attempted to shoot through the roof at them, but to his companions' amusement the bullet did not even penetrate the roof and instead dropped harmlessly to the floor. Thirty years later a company of Arizona miners encountered a burgeoning feline population at the Big Stick Mine. These cats, which had come with earlier mine operators, were for all intents and purposes wild. During the day the animals were rarely visible, but in the evenings they wailed like the banshees of hell. Only a campaign of extermination brought the surplus population into manageable proportions.[13]

In addition to pests, disease menaced the mining towns. The most common ailments were diarrhea and dysentery. Given the quality of the food and the circumstances of its preparation, these complaints are not surprising. While many mining communities had both good restaurants and good supplies of fresh food, the average miner and his family could not afford the high prices of such sustenance. In addition,

[11] George Whitwell Parsons, *The Private Journal of George Whitwell Parsons,* July 4, 1880, pp. 134–135.

[12] "Rats in a Silver Mine," *Mining and Scientific Press* 67 (1893), 123, reprinted from *Engineering Journal.*

[13] Parsons, *Journal,* September 8, 1880, p. 161; Frank A. Crampton, *Deep Enough: A Working Stiff in the Western Mine Camps,* pp. 156–157.

the water supplies were often polluted, affording little escape from amoebic dysentery. One Colorado miner wrote to his wife from his placer operations on the Snake River: "I shall remain here some weeks if I am well but some of us are sick all of the time: We shall get something different to eat." Even Isaac Requa, superintendent of the Potosi Mining Company, was forced to write from Nevada that he was suffering from "looseness of the bowels" and griping pains. In Tombstone, Parsons suffered periodically from similar symptoms; his bouts with diarrhea often lasted several days and kept him bedridden.[14] Yet these were only the minor irritations of life; worse were the seasonal plagues that swept through the towns taking a heavy toll.

Contagious diseases posed serious health hazards on the frontier. Inadequate sanitation, inferior housing, overcrowding, and uncertain access to pure water explain why diseases struck frequently and ferociously in the mining towns. Furthermore, both mining itself and the climate of the region increased miners' susceptibility to respiratory ailments. Finally, mining towns often lacked competent physicians and systematic quarantine procedures.[15]

Nearly every infectious disease found in the United States visited the mining camps of the Rocky Mountain and Great Basin West. A list of these common disorders, often epidemic in proportions, suggests both the seriousness and the contagiousness of the ailments. Careful study indicates the prevalence of diphtheria, pneumonia, typhoid fever, smallpox, influenza, scarlet fever, yellow fever, cerebrospinal meningitis, and less acute afflictions including venereal diseases, common colds, and tuberculosis (often job- or climate-related).

Such afflictions had important consequences for the miners, their families, and their fellow citizens, worst of which was death itself. The seriousness of the disease threat was implied in the pleasure taken by the union miners of Anaconda, Colorado, in reporting on June 10, 1900, that all the members had survived their illnesses of the winter

[14] Wm to Lucia, July 19, 1865, Wm [William] Letters, DPL; Isaac L. Requa to Wife, Virginia City, August 31, 1877, Isaac Lawrence Requa Papers, HL; Parsons, *Journal*, May 31–June 1, 1880, and May 28–May 30, 1881, pp. 127, 231–232.

[15] Ray Colwell, "Cripple Creek," p. 14, Colwell Papers, WHRC; Russell R. Elliott, *Nevada's Twentieth-Century Mining Boom: Tonopah, Goldfield, Ely*, p. 47; F. Sommer Schmidt, "Early Days at the Nevada Consolidated Copper Company," p. 11, Mark Requa Papers, WHRC.

and spring. They proudly noted that "at present we have only one member on the sick list."[16] Even when it was not fatal, serious illness meant that the miner had to curtail his work or stop altogether until he recovered. For a laboring man, inactivity due to physical disorders meant the loss of needed wages and, not infrequently, the loss of his position at the mine.

One resident of Cripple Creek, Colorado (Mabel Lee), spoke of these epidemics as periodic in nature. She recalled that a neighbor's wife had died during a typhoid epidemic in 1902 and the miner-husband had been taken by pneumonia two years later. As a child in 1892, Mabel herself had been ostracized because she and her parents lived in a tent formerly occupied by a family with diphtheria. With only rudimentary medical assistance available, people tried especially hard to avoid contamination. To this end, public places, especially schools, were closed during epidemics. One mine superintendent noted that his entire Arizona operation had been quarantined. "We had to arrange a pest-house, get nurses, and generally do many things to stop the epidemic, and at the same time keep the mine and mill going." He confided that "there were few things I liked less than going into that hot, ill ventilated mine at that time."[17]

In the early twentieth century influenza became the most dreaded affliction on the mining frontier. This was not only due to an increase in the number of influenza cases. Rather, it stemmed in part from gradually improved techniques of treating other, traditional illnesses, especially the increased use of vaccination to control smallpox. In addition, the recognition of influenza's importance grew as doctors were gradually able to identify influenza and separate it from the general rubric, "grippe." Recognized strains of the virus struck the mining camps of Arizona in 1901, and in 1906 it threatened the life of Frank Crampton, then mining in Goldfield. However, the most serious epidemic occurred in the fall of 1918. It ravaged the mining camps of the Rocky Mountain and Great Basin West, where the high altitude compounded it and the isolation of the camps hampered effective control. Among the most seriously affected communities were Mogollon,

[16] Press Committee of Anaconda Miners' Union #21 to Editor, *Miners' Magazine* 1 (July, 1900), 49.

[17] Lee, *Cripple Creek Days*, pp. 23–25, 178–179, 213–214; Staunton, "Memoirs," I, 91, Staunton Papers, UA.

New Mexico, Cripple Creek, Colorado (where about four hundred persons lost their lives), and Silverton, Colorado. In Silverton the epidemic killed more than 150 persons, 25 of whom were members of the local miners' union. The influenza epidemic of 1918 was the last and greatest epidemic to strike this region during this era.[18]

In addition to illness and infestation, miscellaneous accidents posed additional nonoccupational perils. As is their wont, men and women injured themselves by falling from wagons, by stepping into holes or deserted mine shafts, by slipping off ladders, or by tumbling from elevated sidewalks. While by no means exhausting the potential for human ineptitude and carelessness, these examples suggest the types of hazards that have been and are encountered by men, women, and children in their daily lives. For the most part these hazards were not peculiar to the urban settlements of the West; they were and are part of the human experience.

There were diversions as well as discouragements for the miners, although these diversions were not always as colorful as tradition has painted them. The dime-novel lore of the West invariably portrays the importance of saloons, dance halls, gambling dens, and cribs. In the popular mind these establishments have been identified with the leisure time of the Western population, especially its males. Yet contrary to the fictional representation, these establishments were, as miner Leo J. Keena described them, "drab and drunken, and business like, with no nonsense about glamour." All were concerned with turning a profit with the least possible investment. The denizens of these places mirrored their surroundings; the men were seldom chivalrous, and the women, typically unappealing. These varied businesses, catering to a predominantly male clientele, appeared almost everywhere. Only company towns excluded or carefully regulated them.[19]

---

[18] Staunton, "Memoirs," I, 104, Staunton Papers, UA; Crampton, *Deep Enough*, p. 68; E. H. Kennard & Bierce to D. B. Scott, Mogollon [N.M.], November 16, 1918, Blumenthal Papers, Special Collections Department, Zimmerman Library of University of New Mexico, Albuquerque; Lee, *Cripple Creek Days*, pp. 259–261; "Silverton Miners' Union Suffers Severely from the Flu Epidemic," *Miners' Magazine* 19 (December, 1918), 5.

[19] Keena, "Cripple Creek," p. 1, Keena Papers, SHSC; Carl Snyder, "In the Cripple Creek Gold-Fields," *Harper's Weekly* 40 (1896), 59; G. Thomas Ingham, *Digging Gold among the Rockies*, p. 297.

The saloons served as a kind of social center. There men discussed their day's labor, talked local and national politics, and speculated about the future of the mines and their community. They gossiped about the local owners and managers with mixtures of envy, affection, irritation, and derision. Miners tended to gravitate toward particular bars. They might join their fellow nationals in French, Italian, Cornish, or Irish bars; they might accompany the men of their shift to a nearby pub; or they might drink with men from their neighborhood. They found smoke-filled rooms, rotgut whiskey, watery beer, and human contact.[20]

The lure of the gaming palaces with their draw and stud poker games, keno, roulette, and faro was equally strong. According to John T. Waldorf, gambling was the endemic pastime of the mining frontier; speculation of one kind or another attracted almost everybody. "The babies were barred, of course, but I've seen infants making a faro chip do duty as a teething ring. We small boys went further and played cards for matches. Sometimes we got blooded and bet such treasures as tops and marbles. . . . The favorite game with our fathers was stocks." Gambling was exciting; each roll of the die, flip of a card, or spin of a wheel beckoned the hopeful. In a society where everyone gambled with nature, anything seemed possible. Yet this optimism was countered by the certainty that there was little hope of beating the odds. In Utah the miners' unions even joined forces with the Mormon establishment to enact legislation that banned gambling.[21]

Stock speculation was a particularly infectious form of gaming in communities like Virginia City. At least one Comstock miner, James Galloway, left California for Virginia City in order to speculate in Nevada mining stocks. He took work in the mines to support himself and his family, as well as to profit from inside information. Four years and hundreds of shifts later, an older but no wiser Galloway wrote, "After 2 years of struggle on the Pacific Coast I am a poor man, but

[20] Theodore F. Van Wagenen, "Colorado," pp. 36–37, Theodore F. Van Wagenen Papers, DPL; J. C. Kinnear, "Fifty Years in Mining, 1907–1957," p. 7, J. C. Kinnear Papers, WHRC; Keena, "Cripple Creek," pp. 5–6, Keena Papers, SHSC.

[21] Waldorf, *Kid on Comstock*, p. 130; Vernon H. Jensen, *Heritage of Conflict: Labor Relations in the Non-Ferrous Metal Industry up to 1930*, pp. 263–264.

with a conscience clear that I am an honest man." It was not the gaming tables, but rather the stock certificates, the illusive hope of a lucky break, that pulled him into continual speculation.[22]

Closely linked to the bars and dance halls, at least initially, were the frontier theaters. Regularity in theatrical performances was a luxury only the larger communities could afford; however, many lesser camps were sporadically visited by traveling vaudeville, minstrel, and thespian companies, lecturers, humorists (including the renowned Artemus Ward), and an assortment of nondescript performers. Respectable alternatives to saloons were widely applauded by community boosters, but this did not ensure a warm reception. Success stemmed from an understanding and appreciative audience, as the Gruenwalds, John Langrishe's operatic stars, discovered when they visited Colorado in 1864. Appearing first in Denver, self-acclaimed cultural center of the Rockies, they aroused a storm of controversy from an unappreciative audience; however, a special request by the residents of Central City brought the Gruenwalds to the mountains and a more sympathetic patronage. One historian of Colorado's theater suggests that Central City's cosmopolitan population, especially the Cornish, Italian, and German contingents (predominantly miners), had been exposed to classical music and opera in their homelands and thus savored what their provincial neighbors in Denver scorned.[23]

The theatrical fare available to miners varied considerably. At one extreme were the classic dramas of the English theater, specifically those of Shakespeare and Marlowe, while other presentations were more aptly described as variety acts and circus performances. Occasionally audiences witnessed a reading of *Hamlet*, scenes from popular nineteenth-century comedy like *Our American Cousin*, slapstick, and tightrope walking or acrobatics on successive evenings. Sometimes all shared the stage on a given night.[24]

For miners and their families the theater provided unusual sights, excitement, laughter, and brief escapes from the harsh and dreary

[22] Galloway, *Engineering Works*, pp. 15, 18–20.

[23] Melvin Schoberlin, *From Candles to Footlights: A Biography of the Pike's Peak Theatre, 1859–1876*, pp. 133–136; Lynn I. Perrigo, "The First Two Decades of Central City Theatricals," *Colorado Magazine* 11 (July, 1934), 141–152.

[24] Wells Drury, *An Editor on the Comstock Lode*, pp. 54–59.

atmosphere of the towns. Children enjoyed the productions as much as their parents, according to John Waldorf, who with a friend had once entered the skylight of Virginia City's Piper Opera House in order to see the famed dueling scene in Shakespeare's *Hamlet*. Working-class families formed an appreciative, though critical, audience for the traveling companies. Fans occasionally showered their favorite actors and actresses with gold and silver coins or gave them special benefit performances; westerners rewarded the veneer of culture that the shows provided.[25]

Another important pastime was reading, anything from novels and scientific texts to letters from home. Newspapers appeared in most communities, and miners often subscribed to papers or periodicals from back home. Financed by subscriptions, wealthy individuals, companies, public assessments, fraternal organizations, or miners' unions, local libraries sprang up. Virginia City's union maintained a members' library containing several thousand volumes. Young John Waldorf read widely in this collection and found in it the basis for a journalist's career. Literary societies and even union locals provided forums for discussions of literature, politics, religion, and philosophy. Even the *Miners' Magazine* published poetry and a reading list of mining fiction.[26]

Sedentary card games appealed to some men, as did checkers. These amusements spawned companionship and friendly competition. Both men and women gathered to listen to the yarns of storytellers, to relax around stoves and bonfires, and to laugh and cry over the human condition. Courting and dancing were other outlets for sociability.

[25] Waldorf, *Kid on Comstock*, pp. 62, 149–150; Galloway, *Engineering Works*, p. 20.

[26] Lord, *Comstock Miners*, pp. 212–215, 377; Galloway, *Engineering Works*, p. 20; Waldorf, *Kid on Comstock*, pp. 7–8; Altman Stationary Engineer's Press Committee to Editor, *Miners' Magazine* 1 (April, 1900), 21–22. Miners' verse appeared regularly in the *Miners' Magazine*, and in 1901 a poem was used to nominate Ed Boyce for the presidency of the Western Federation of Miners (*Official Proceedings of the Ninth Annual Convention of the Western Federation of Miners of America, Denver, Colorado, May 27–June 6, 1901* [Pueblo, Colorado: Pueblo Courier Print, 1901], p. 101). The bibliography of mining fictions was compiled by C. D. Spivak, M.D., and appeared in *Miners' Magazine* 1 (April, 1900), 18–19. See also Diary of Albert B. Sanford, June 4, 1881, Albert Byron Sanford Papers, SHSC; Diary of E. D. Gardner, May 5, 1907, E. D. Gardner Papers, WHRC.

For the bachelor miner, feminine company was a sweet diversion or, as one young man remarked, the source of a "most pleasant evening." Consequently, lodges, churches, clubs, and unions hosted Labor Day balls and Christmas parties where men and women mingled.[27]

Then as now, some men relaxed by admiring the spectacular natural beauty of the region. Mountains and meadows, aspen groves and fields of columbines, wild berries and scampering wildlife, canyons and gorges, all of these and more drew miners and their families into the countryside for an occasional outing. Here they found breathtaking splendor and secluded spooning spots. Families enjoyed picnics in the country, and unions customarily sponsored one great picnic per year for their membership. In Leadville, it was the most important event that the Cloud City Miners' Union sponsored. Requiring careful planning and preparation, it included races and contests, in addition to good food and congenial company.[28]

In the mining communities as elsewhere, the American male's need for companionship and camaraderie found expression in the proliferation of fraternal organizations. Ray Colwell explained, "Partly, I suspect, because there were so many single men who had time on their hands in the evenings, all the well known fraternal orders were represented, as well as many that have long since vanished." Among the most common groups were the Ancient Order of United Workmen, the Masons, the Eagles, the Knights of Pythias, the Odd Fellows, and later the Elks. In addition, the fire-fighting companies were semifraternal organizations. That the reality of membership in fraternal groups often involved bickering and infighting did not deter men from joining. Some miners were attracted to the fraternal organizations not only for the fellowship but also because they offered death benefits. Though the sums were not large, they were the only source available

[27] Bessie Launder Richards, "Mining Town Memories—Colorado and Mexico," typescript of interview conducted by Mel Erskine, p. 38, BL; Parsons, *Journal*, February 19, 23, and 24, 1880, pp. 93, 95, 96; Diary of Sanford, July 29, 1881, Sanford Papers, SHSC, Journal of Alfred P. Camp, July 27, 1875, Alfred P. Camp Papers, SHSC; Report of Operative "P. B." [C. L. Knuckey], July 20, August 10, November 23, and November 30, 1900, Campion Papers, CU.

[28] Diaries of Major O. S. Alers, 1901–1909, O. S. Alers Papers, WHRC; Report of W. E. Giese's Operative in Leadville, June 29, 1899, and Report of P. B., August 21, August 31, and September 21, 1900, Campion Papers, CU.

to miners, who were unable to purchase ordinary life insurance. The miners hoped that the lodge benefits would be enough to pay for their burial.[29]

One event that brought pleasure to all was the arrival of a circus. Circuses had special appeal for the young, who laughed at the clowns, gaped at misfits and wild beasts, and shuddered at the sight of tight-rope walkers and aerialists. Schools closed when a circus came to town, and youngsters rose at the crack of dawn to watch animals unloaded and the big top raised. Everyone noted the circus's arrival and reveled in the escape it offered.[30]

Important community events provided additional diversions from the miner's daily routine. Entire communities toasted the completion of important public buildings, the inauguration of regular mail service, the activation of waterworks and sewage facilities, or the arrival of telephone or telegraph systems. Railroads were absolutely essential to the growth and development of a mining district, and their arrival in a mining town invariably precipitated widespread celebration.[31]

Typically, the first train to reach the station was met by an enthusiastic crowd, which proclaimed a holiday. C. H. Hanington remembered Central City's first train as an occasion for "another civic celebration including a welcoming committee of fire companies, fraternal societies, bands, the citizenry and the Emmet Guards." Similar excitement swept Virginia City in 1869 when the Virginia and Truckee Railroad linked it to the Central Pacific and San Francisco. Tombstone celebrated with horse races and a schedule of gala events.[32]

An odd assortment of holidays and events brought additional diversions. For example, on March 11, 1895, the miners of Delamar,

<hr />

[29] Ray Colwell, "Cripple Creek," p. 21, Colwell Papers, WHRC; Elliott, *Nevada's Boom*, pp. 49–51, 240–245; Roll Book of Members, Select Knights of Virginia City, Ancient Order of United Workmen, Virginia City Lodge No. 1, Papers, BL.

[30] E. Louise Presser, "Memoirs of E. Louise Presser, Rhyolite, 1906–1908," *Nevada Historical Society Quarterly* 15 (Summer, 1972), 48–51; Waldorf, *Kid on Comstock*, pp. 140–141.

[31] C. H. Hanington, "Early Days of Central City," *Colorado Magazine* 19 (January, 1942), 12; Lee, *Cripple Creek Days*, pp. 100–101; Albert S. Frost to Colonel R. C. Getch, Cripple Creek, January 14, 1901, Albert S. Frost Papers, SHSC.

[32] Hanington, "Central City," p. 13; William Wright, *The Big Bonanza*, pp. 165–167; Parsons, *Journal*, March 20, 1880, p. 105.

Nevada, celebrated that district's original gold discovery, and Cripple Creekers honored the creation of Teller County with special festivities on March 23, 1899. Tragedy brought grieving or merely curious miners to the funerals of local dignitaries, notorious gamblers and prostitutes, and fallen comrades. Even the departure of close friends or the breakup of an underground team of miners occasioned one last fling.[33]

A number of special national events drew the mining camps into a shared sense of exultation or mourning. Miners reacted corporately to the assault on Fort Sumter, Lee's surrender, the assassinations of Lincoln, Garfield, and McKinley, and the declarations of war in 1898 and 1917. These events were of such profound national importance that the Westerners were deeply moved and felt a kinship with other Americans. Being a diverse population, they also observed national and religious holidays of various origins. In the Southwest, Mexican-American miners participated in their society's traditional feast days and festivals. W. Turrentine Jackson found that St. Patrick's Day, Washington's Birthday, and Thanksgiving were among the traditional holidays of the White Pine District. Thanksgiving, then as now, was primarily a day of rest and feasting.[34]

The Christmas–New Year's celebration was the most sentimental of the mining-town holidays, as residents reflected upon other times and other places. Traditionally, Christmas had been a family celebration; the absence of familiar friends and relatives made the season sad and nostalgic, especially in isolated and remote mining towns. Writing home from the Colorado placer camp of Spanish Bar in 1861, Frank Hall gave expression to the thoughts of thousands who followed him. "Christmas passed off quietly enough. After partaking of a hearty breakfast, I sat down, poked my feet up on the chimney side, and smoked my cares away. In dreams my fancy revelled of the past[,] the present, with an occasional stray glimpse toward the future." Yet withal they made the best of it. They planned dances, erected Christ-

---

[33] Townley, "Delamar Boom," p. 9; Lee, *Cripple Creek Days*, pp. 80–84, 138–147; Guy Mayo, "Rawhide Pays Its Respects: Wherein Riley Grannan, King of the Sport of Kings, Runs His Last Furlong," *American West* 10 (January, 1973), 41; Crampton, *Deep Enough*, p. 46.

[34] Thomas M. Yenkes to Henry D. Bacon [New Mexico], Henry Douglas Bacon Papers, HL; W. Turrentine Jackson, *Treasure Hill: Portrait of a Silver Mining Camp*, pp. 78–79.

mas trees, and exuded a warm spirit of conviviality. In Prescott saloon-keepers and gamblers hung cast-off garments and boots upon juniper and cedar trees like socks upon the chimney; everything was appropriately labeled and contained a pint or quart of the "Christmas spirit." Elsewhere greenery, mistletoe, and lights adorned saloons, hotels, and even the cribs. It was a decorative, festive affair, but one tinged with nostalgia.[35]

Independence Day has always been a symbolic invocation of American nationalism, and out west it became the most important occasion of each year. Work ceased on the Fourth and only gradually resumed over the next several days. As John T. Waldorf remembered it, observance of the holiday began before the crack of dawn. "Fourth of July and 'don't forget it' which we did not when the night shift after retiring at 3 o'clock from work let off a royal salute on the surface with giant powder which aroused all hands." In a society where most men knew something about powder, Independence Day invariably arrived and departed with a "bang!" Youngsters detonated their own explosives: redheads, bombs, and double-headed Dutchmen, to name but a few. Children and mature bullies made life miserable for peace-loving folk; firecrackers sailed from every direction, frightening horses and irritating, if not injuring, the unsuspecting.[36] Francis Scott Key would have found ample inspiration for "The Star-Spangled Banner" during the typical Independence Day celebration on the mining frontier.

Activities included one or more patriotic readings or addresses. Various contests encouraged participation by the adventurous and provided spectator sport for the others. Foot races were common, as were horse races; wrestling contests were popular, and sparring matches also attracted attention. Dog fights, cock fights, bear-bull fights, and other vicious entertainment occasionally took place. Dances and picnics were invariably parts of the celebration. There were contests that developed directly from daily life, notably mule-packing and

[35] Frank Hall to Mother, January 6, 1861, Hall Papers, DPL; Thomas D. Sanders and Charles C. Stemmer, "Hardships and Perils on the Western Frontier or Fifty Years of Pioneering," pp. 117–118.

[36] Parsons, *Journal*, July 4–July 6, 1880, pp. 134–136; Waldorf, *Kid on Comstock*, pp. 98–100; Robert Livermore, "An Autobiography," p. 126, Robert Livermore Papers, WHRC.

rock-drilling contests. By the turn of the century these events had achieved great popularity and attracted semiprofessional competitors who contended for hundreds of dollars in prize money. Finally, the Fourth, like all mining town holidays, meant additional business for the saloons, dance halls, gambling parlors, and cribs. Throughout this era it was a raucous and exciting summer festival.[37]

In 1894 Congress made Labor Day a national holiday, and from the beginning miners associated it with the formation of the Western Federation of Miners (WFM), organized in 1893. This perceived link meant that WFM locals planned and promoted the Labor Day festivities. Their celebrations usually involved closing the mines and included activities comparable to those of the Fourth of July, minus the fireworks and intense patriotism.[38]

Even when there was no holiday, miners and their families enjoyed physical activity. As noted, sporting events were popular during celebrations. However, miners and their children engaged in various forms of physical recreation throughout the year. In the hilly environment of most towns, they enjoyed winter coasting, skiing, and tobogganing. Mabel Barbee Lee remembered that children used "every imaginable device—barrel staves, scraps of sheet iron, dishpans and homemade sleds." The heavy snows of Colorado's western slope provided excellent skiing and snowshoeing. In Central City the basement of a gutted hardware store served as a ready-made ice skating rink. Even some indoor winter activities were athletic in nature; roller skating had become popular in Colorado by 1885.[39]

In spring, summer, and fall children mixed games of imitation and fantasy with such conventional ones as hopscotch, jacks, kick-the-can, tag, jumping and climbing rope, stilt-walking, marbles, and swimming. They lived in a rich and varied physical environment,

[37] Galloway, *Engineering Works*, p. 20; Wm to Lucia, Nevada City, July 4, 1865, Wm [William] Letters, DPL; "The Fourth of July in Cripple Creek," *Miners' Magazine* 1 (August, 1900), 39; Kinnear, "Mining," pp. 9–10, Kinnear Papers, WHRC; Richards, "Memories," pp. 39–40.

[38] Diary of Gardner, June, 1907, Gardner Papers, WHRC.

[39] Lee, *Cripple Creek Days*, p. 39; Galloway, *Engineering Works*, p. 20; Waldorf, *Kid on Comstock*, pp. 2–3; Harry C. Cornwall, "My First Year in the Gunnison Country," ed. Duane Vandenbusche, *Colorado Magazine* 46 (Summer, 1969), 231–234; Hanington, "Central City," p. 14; "One Way of Living," *Denver Rocky Mountain News*, July 6, 1891.

which, though often dangerous, offered opportunities for investigation, exploration, and adventure of the derring-do variety. In addition, they joined their parents as spectators of adult sporting events. Racing of anything and everything, horses, mules, dogs, turtles, rabbits, wagons, and eventually cars, was popular.[40]

Hunting and shooting matches provided sport and competition for miners so inclined. Fishing and hunting also brought variety to their diets and helped to reduce the cost of living. Additional recreation was offered by organized athletic contests, which had followed the men to the camps. By the mid 1860's baseball fever had swept the mining camps of Colorado and Nevada. Neighboring mining towns often proved their worth with a "friendly" game of baseball. C. H. Hanington reported that these struggles occasionally ended in a wild free-for-all of spectators, umpires, and players. Football, basketball, tennis, and bowling were brought west later by young college men, and the popularity of these sports spread in the twentieth century.[41]

Other than drilling contests, wrestling and boxing were the miners' favorite sports. Wrestling matches pitted brute strength against brute strength, mental agility against mental agility. The Cornishmen especially loved their amateur wrestling matches and promoted competition among themselves. Boxing had an even larger following. Ranging from simple brawls to national matches, fighting provided exciting fare for the spectators, ordinarily a cash prize for the contestants, and a chance for all to wager on the outcome. Boxing appeared early on the mining frontier, when enterprising promoters with their pugilists in tow offered sparring exhibitions or gave the local talent a chance to go a round with the "champ." Prize fighting was popular with all classes and age groups. Lowell Thomas, in a foreword to Mabel Lee's book, recalled that he and his Cripple Creek companions followed the careers of John L. Sullivan, Gentleman Jim Corbett, Jim Burns, Stanley Ketchel, Joe Gans, and Ad Wolgast.[42]

[40] Lee, *Cripple Creek Days*, pp. 45–46, 111–123; Waldorf, *Kid on Comstock*, pp. 52–53, 106–108; Lynn I. Perrigo, "The Cornish Miners of Early Gilpin County," *Colorado Magazine* 14 (May, 1937), 97.

[41] Livermore, "Autobiography," p. 86, Livermore Papers, WHRC; John Marshall Paul to Little John, Denver, November 3, 1866, John Marshall Paul Papers, SHSC; Diary of Arthur Bowen, June 3 and June 4, 1882, DPL; Hanington, "Central City," p. 14; Elliott, *Nevada's Boom*, pp. 55–56, 240–245.

[42] Perrigo, "Cornish Miners," p. 97; Hanington, "Central City," p. 8;

Drilling contests were the miners' own distinctive sporting events. Analogous to the cattle range's rodeo and the Midwestern fair's plowing and pulling contests, they attained their greatest popularity in the states and territories of the Trans-Mississippi mining frontier. Like their plains and prairie counterparts, drilling bouts became prominent at the very time that the skills they glorified were declining in importance. These contests were part of an earlier era of sport in which vocational prowess provided recreational outlets.

Drilling contests were invariably held during the summer months and customarily played a prominent part in the festivities organized to celebrate Independence Day and Labor Day, though they were also scheduled for special occasions like Tonopah's Railroad Days. They were held in summer to guarantee that interested contestants and spectators could attend them with no interference from the weather; holidays were chosen so that work did not hinder anyone. The contests pitted individuals or teams of two miners against one another. They centered on hand-drilling, an essential aspect of the hard-rock miner's work. Using hammers and steel drills, laboring miners punched holes into the underground rock, filled them with explosives, and detonated these charges, breaking the rock and ore from the facing. Drilling contests tested the participants' skill at this particular task; the men bored into a boulder, usually Colorado or Vermont granite, over a specified period of time, commonly fifteen minutes. If done by one individual holding his own drill bits and hammering with his other hand, the drilling was referred to as single-jacking; if performed by a two-man team with one holding the bits and the other wielding a heavier hammer it was called double-jacking.

Whether single- or double-jacking, the competition required courage, skill, strength, and endurance. The two types of drilling were frequently scheduled on successive days so that men might compete in both categories. Prize money ranging from several hundred dollars to several thousand dollars proved an added incentive. The winners and runners-up were the sole recipients of money, but local businessmen and mining company officials provided other prizes: loving cups, silverware, clothing, drills, and hammers. In addition to the physical

Elliott, *Nevada's Boom*, pp. 55–56, 240–243; Lee, *Cripple Creek Days*, pp. xii–xiii.

prowess required, a successful performance demanded careful preparation of equipment. Drills were carefully sharpened, and their length was carefully measured. Drills were exchanged and replaced as the holes grew deeper. The miners sought to keep the drills in place as long as they retained their edge; therefore, contestants whose drills held their edge and were slightly longer had an advantage over others who lost time changing drills. Hammer weight was also important, as was the shape of the hammer's striking face. Miners wanted to bring all their force to bear on the drill heads, which were about an inch in diameter.

The selection of one's drilling location was equally important. A practiced eye could determine softer spots in the granite boulder; the choice of where to drill meant the difference between winning and joining the also-rans. Timing of the hammer stroke and careful drill placement were also essential. A missed stroke might mean a broken arm for a contestant in the single-jacking and almost certainly did for his companion in the double-jacking. The heavier hammer and the two-handed blow of double-jacking left no room for carelessness on the part of the drill holder. If not held perfectly straight, the drill would bind or "fitcher" and would require tapping to loosen it. Occasionally drills were bound so tightly in the rock that the team had to begin another hole. In either case precious time was lost, and the man or men could not expect to win.

Both single- and double-jacking required careful pacing of the contestants' effort. Every stroke of the hammer had to be precisely delivered; the first few strokes determined whether the hole would be straight; the remainder required every ounce of strength of the participants. Single-jacking required careful marshalling of stamina; double-jacking not only required the same attention to stamina but also involved the men's changing positions every thirty seconds. A smooth transition from hammer-holder to drill-holder was essential; good teams made the transition without the loss of a single stroke. Each man had his own drills and hammer to cut confusion to a minimum.[43]

[43] The literature on rock drilling is scattered but fairly extensive. The summary account of contests presented here relies primarily upon the following: Brown, *Lady in Boomtown*, pp. 73–75; Crampton, *Deep Enough*, pp. 47–48, 56–58; Victor I. Noxon, "Hard Rock Drilling Contests in Colorado," transcribed and prepared by Forest Crossen, *Colorado Magazine* 11 (May, 1934),

Drilling contests provided exciting entertainment for the crowd of spectators. Each team had its own interested followers who cheered their every exertion. Betting was brisk, which added to the enthusiasm of the audience. Cheers, catcalls, oaths, and screamed incentives were directed toward the participants; the scene was one of enthusiastic bedlam. Mrs. Hugh Brown later recalled the commotion she had heard during Tonopah's drilling contest in 1904: "Lean on that stick there!" "Two to one on this team!" "We got our money on ya, Bill!" "I'll take ya!" "Put 'er down, boys!" For the winners there was a warm reception; they were recognized as champions, applauded, and feted wherever they went.[44]

When drilling contests attained their greatest popularity, in the early years of the twentieth century, the skills perfected were already giving way in the mines to increased industrialization and technological innovation. By 1914 the old drilling contests with their emphasis on single- and double-jacking had largely disappeared. Insofar as drilling contests survived technological changes in mining techniques, they did so as remnants of the past. Increasingly the contests utilized mechanical drills and pitted only local drill teams against one another. According to onetime miner Albert Frantz the contests lost much of the enthusiasm that had attended the turn-of-the-century affairs. As machines assumed an ever more important role in mining, the previous pride in individual accomplishments languished.[45]

In the predominantly masculine society of the mining frontier, various kinds of "initiation rites" provided amusement and distraction for the working men. Newlyweds were often given their first taste of community life with a rousing shivaree, a raucous clamoring and clattering of tin cans, bells, horns, and any other noisemaker the groom's friends could find. Dudes or greenhorns were the objects of the most interesting and sometimes cruel forms of frontier humor. The newcomer, particularly one who fit the miners' stereotype of an

81–85; "Our Champion Miners," *Denver Rocky Mountain News*, October 19, 1895; "Rules and Prizes for Miners' Drilling Contest," *Denver Rocky Mountain News*, August 14, 1899; "Drilling Contest at Tonopah," *Miners' Magazine* 18 (October, 1917), 4; Telegram, Kenneth L. Fahnestock to John F. Campion, Leadville, October 23, 1901, Campion Papers, CU.

[44] Brown, *Lady in Boomtown*, p. 74.

[45] Interview with Albert J. Frantz, Denver, April 6, 1972.

eastern dandy, was certain to attract practical jokers. In the depths of the mines these dudes might be sent on a series of hopeless quests for imaginery tools like "shaft stretchers," "sky hooks," and "left-handed wrenches." Often, concerted efforts were made to humiliate the newcomer in front of the entire community. Snipe hunts were a favorite source of amusement. Usually only humiliating, this prank could become dangerous in the mountains, where dangers from nature and wildlife lurked in the dark. The unsuspecting tenderfoot was guided into the hills by his companions and then left in a desolate spot to wait for the "dangerous" (nonexistent) snipe. At this point the pranksters returned to town and awaited the arrival of the frightened and bedraggled, but wiser, victim.[46]

These practical jokes were tame compared to the enthusiasm that attended so-called badger fights, the ultimate caper. Badger fights were usually reserved for gullible and self-assertive fellows. A successful badger fight required the acquiescence of the entire community. Preparations usually began with an offhand reference to badger-versus-dog fights. If the newcomer expressed ignorance of such affairs, the serious planning began. The principals in the projected fracas were the naive wretch, a ferocious dog, and the "vicious and fearless" badger (really a metal chamber pot). First the miners organized parties to hunt for and capture a wily badger. While these early preparations might involve the greenhorn, he was never permitted to see the badger. Finally, on the day of the "big fight" the beguiled stranger would be designated to release the badger from its cage. This was supposed to be done with the assistance of a long rope attached to the cage; unbeknown to the courageous stranger, however, the rope was also firmly attached to the "badger." When the signal was given, the frightened fellow jerked the rope and discovered that the "badger" was attached and in hot pursuit. When he discovered the ruse, usually after having frantically tried to escape the ostensibly vicious creature, the victim either took the torment in good spirits or responded by threatening violence and abusing the crowd with profanity. The good sports were recognized as "regular" fellows; the hotheads merely added to the humor of the pranksters. At best the hotheads responded

[46] Brown, *Lady in Boomtown*, p. 30; Lee, *Cripple Creek Days*, p. 250. The greenhorn stunts are described in Crampton, *Deep Enough*, pp. 122–123, 212–213.

as had been expected, and their status in the community did not improve.[47]

Other practical jokes appeared from time to time. Once Frank Crampton, having won a bag of small change at the roulette wheel, offered to buy drinks for the assemblage from his winnings. He asked the bartender if he would accept some of this "chicken feed" in payment. When the latter agreed, he turned over a portion of his coins. Another fellow offered to buy this lucky "chicken feed" and then returned to the roulette table. Like Crampton, he too won at roulette and offered to buy drinks for everyone. He too asked the bartender if he might pay in "chicken feed." The bartender replied in the affirmative, and the man proceeded to reach into his pocket and pull out a handful of cracked corn. Though initially furious, the bartender quickly accepted the proffered corn, and the joke was accepted in good grace. Crampton noted that thereafter the "chicken feed" was prominently displayed beneath a carefully lettered sign which said: "We do not need chicken feed. We have more than we can use."[48] Here was a unique example of the miners' penchant for practical jokes and the willingness of their fellow citizens to abide by the rough-and-ready humor of the mining frontier.

There was a vast emotional chasm between the raucous laughter of miners as pranksters and their bitter complaints as victims of fire or of the recurrent plagues: smallpox, cholera, diphtheria, and influenza. Both experiences were part of the miner's world. Fire, lawlessness, diseases, pests, and accidents threatened the miners and their communities, but they did not keep these people from enjoying themselves, with diversions that ranged from the saloons and dance halls to the special holidays and recreational sports characteristic of the mining west. Above ground the miner was friend, father, husband, community booster, and an ordinary fellow, but underground he bore the sobriquet "hard-rock stiff" and faced all the problems that attended his calling.

[47] Crampton, *Deep Enough*, pp. 123–130; Elliott, *Nevada's Boom*, pp. 243–244.

[48] Crampton, *Deep Enough*, pp. 135–138.

4.

At Work amidst the Ore

Clink! Clink! Clink!
The song of the hammer and drill!
At the sound of the whistle so shrill and clear,
He must leave the wife and the children dear,
In the cabin upon the hill.[1]

THIS simple verse was written by a miner who knew both the exhilaration and inherent dangers of his calling. Cornishman Alfred Castner King had turned to poetry when a premature explosion robbed him of his sight. Eulogizing the miners, he understood their firm resolve as they made their way to the shaft houses and collars. He had known the hard work of miners and had experienced one of their debilitations.[2]

Miners lived and reveled on the surface; they worked in the subterranean depths, in the shafts, drifts, levels, and stopes of thousands of mines. The terminology of underground work itself deserves brief comment. In general mining parlance, a shaft is a vertical or inclined hole into the ground, providing access to the ore deposits or veins through a collar or opening at the surface. It is through the shafts that men and material are transported into and out of the ground. Drifts are horizontal tunnels that follow the mineral veins, while levels are all the passageways on a given horizontal plane of the mine workings. Typically, levels occur at regular intervals, are numbered in hundreds of feet from the surface, and provide access to the vein. Technical distinctions aside, the terms *drift* and *level* were often interchangeable in the miner's vernacular. Like drifts and levels, a crosscut is a horizontal passageway driven at right angles to a drift or level, or (as in Figure 2) across, towards a vein or ore body. Stopes are excavations

[1] Alfred Castner King, "The Miner," *Mountain Idylls and Other Poems,* p. 46.
[2] Ibid., pp. 9–10.

at the vein or ore deposit and are the holes through which ore is initially removed. Stopes extend above or below the level; ore removal above the level is called overhand stoping, that below, underhand stoping. Simple schematic sketches of mines with passageways and stopes labeled appear in Figures 1 and 2.[3]

In bonanza times the working miners labored around the clock in underground works that knew neither night nor day, only eternal darkness broken occasionally by artificial lighting. In the 1860's the first shift or work period began promptly at 7:00 A.M. One notable exception to this early starting time was at the Seven-thirty Mine of Brownsville, Colorado, so named because its first shift began at 7:30 A.M. to permit commuters from nearby Georgetown and Silver Plume additional time.[4] The start of the shift meant that the miner was to be at his appointed task; only in the twentieth century did it come to designate the time at which the miner reached the shaft or tunnel entrance.

For the miners working this early shift, preparations began as much as two hours earlier. First, the men had to get dressed. Typically they wore some combination of "rough circular jackets or ragged coats stained and smeared with clay, loose woolen shirts, blue duck overalls wrinkled and dirty, heavy shapeless brogans, and coarse felt hats," all spattered with candle drippings. In winter they added buckskin gloves, leather jackets, and primitive sunglasses that broke the wind and protected their eyes from blowing snow or bright glare. Their apparel was practical and cheap, suited to the hard, dirty work.[5]

[3] Definitions of mining terms are drawn from Albert H. Fay, "A Glossary of the Mining and Mineral Industry," U.S. Bureau of Mines, *Bulletin 95* (1920). Of more limited value but more recent issue is Otis E. Young, Jr., *Western Mining: An Informal Account of the Precious-Metals Prospecting, Placering, Lode Mining, and Milling of the American Frontier from Spanish Times to 1893*, especially Appendix 2, pp. 293–317.

[4] George Rowe, "The True Story of a Rich Silver Mine," George Rowe Papers, Western History Department, Denver Public Library (hereafter cited as DPL).

[5] Eliot Lord, *Comstock Mining and Miners*, p. 312; Leo J. Keena, "Cripple Creek in 1900," p. 1, Leo J. Keena Papers, State Historical Society of Colorado, Denver (hereafter cited as SHSC); William Wright, *The Big Bonanza*, p. 222. In the early twentieth century safety hats began to replace the sloppy felt hats, and special boots with reinforced steel toes supplanted earlier footwear. By the late 1920's Arizona's copper miners were required to wear this new safety apparel (E. D. Gardner and D. J. Parker, "Safety Organiza-

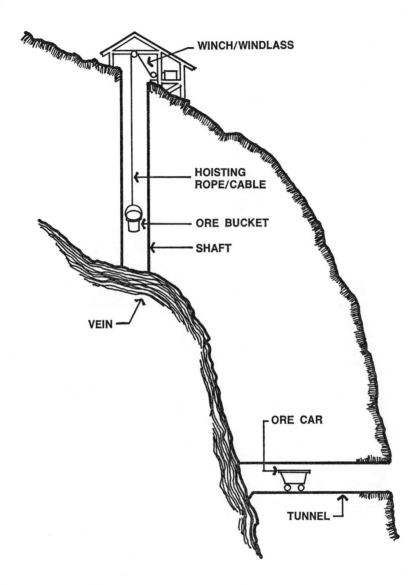

Figure 1. Sketch of Small Mine Type

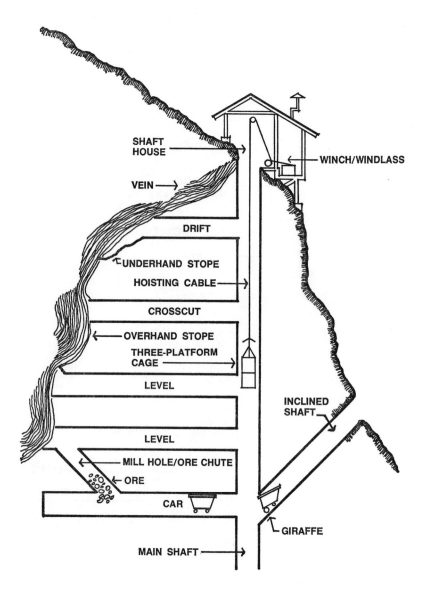

Figure 2. Sketch of "Extractory" Mine Type

Once dressed, the miners gulped down a quick breakfast at home or at a nearby restaurant. Before trudging off, they grasped their last two accoutrements, a lantern or candle and their trusty, tin lunch pail. Ordinarily they walked or, when late, trotted or even ran to work. They traveled from a few blocks to several miles to reach the mines. In larger communities with "suburbs," men often commuted greater distances, as, for example, from Victor or Independence to Cripple Creek. Tombstone's George Parsons, who claimed to have walked about ten miles round trip for work, complained of the "weary—weary time of it I have morning and evening with that tramp—tramp—tramp and hard-hard-labor through the day." The cities often had public transportation—horse-drawn buses, electric railways, or street cars—but elsewhere there were only horses or burros.[6]

Varying from place to place and changing over time, the mines' work shifts ranged from eight to twelve hours in length. Like wages, the work span depended on the prosperity of the mining operations, the presence or absence of a strong union, state legislation governing the conditions of labor, and local practice. For example, as early as 1866, Virginia City miners obtained the eight-hour day. By the early twentieth century the combination of strong regional unionism, under the aegis of the Western Federation of Miners (WFM), and state eight-hour laws had brought the eight-hour day to Cripple Creek, the principal Utah camps, and the Nevada towns of Goldfield, Tonopah, and Ely. However, hours were never effectively standardized. Miners in Bingham Canyon reportedly were still working ten-hour shifts as late as 1917. Some mines operated with two ten-hour shifts, while others retained twelve-hour shifts. A few mines like Leadville's Ibex mixed their work periods; the Ibex had the standard eight-hour shifts,

---

tions in Arizona Copper Mines," U.S. Bureau of Mines, *Technical Papers*, 452 [1929], 21–22, 29, 31–32, 33, 37–38).

[6] Ross B. Moudy, "The Story of a Cripple Creek Miner," *Independent* 57 (August 18, 1904), 382; Lord, *Comstock Miners*, p. 312; Keena, "Cripple Creek," p. 1, Keena Papers, SHSC (Keena described the miner's lantern as "a glass tube with a candle inside and a wire handle"); George Whitwell Parsons, *The Private Journal of George Whitwell Parsons*, March 27, April 5, and April 29, 1880, pp. 107, 111, 117.

a special ten-hour shift beginning at 5:30 P.M., and twelve-hour shifts running from 7:00 to 7:00.[7]

Customarily the miners changed shifts every few weeks. Since there was rarely a wage differential, all worked the irregular or unpleasant shifts. Producing mines operated 6, or more often, 7 days per week. Theoretically miners could expect to work fifty-two weeks out of the year and as many as 362 days. However, as R. Orchard Old noted, work stoppages, occasional holidays, Sundays, and funerals reduced the actual number of workdays to 285.[8]

There neither was nor is such a thing as a typical mine. Mines varied in the quality and quantity of their ores, in their geological characteristics, in their location, and in the extent of their underground workings. The western mines ranged in size from "glorified prospect holes" employing one or two men to the vast subterranean "extractories" employing hundreds of men. The small mine was everywhere; initially all mines began as prospects employing only a few men. However, small mines persisted only in marginal areas.[9]

In the bonanza mining districts, small mines were gradually acquired by the larger corporations or were consolidated with other small holdings to promote operating efficiency and to eliminate litigation or the threat of litigation, which often amounted to financial blackmail. These large operations often employed hundreds of men performing specialized tasks. The holding might include not only the mine proper, but also rail spurs, blacksmith shops, orehouses and

[7] Seth Cook to J. H. Brewer, Virginia City, January 23, 1863, Sierra Nevada Silver Mining Papers, Huntington Library, San Marino, California; Lord, *Comstock Miners*, p. 225; Moudy, "Cripple Creek Miner," p. 382; F. Sommer Schmidt, "Early Days at the Nevada Consolidated Copper Company," pp. 11–12, Mark Requa Papers, Western History Research Center, University of Wyoming, Laramie (hereafter cited as WHRC); "The Miners of Bingham Canyon, Utah," *Miners' Magazine* 18 (May, 1917), 5, reprinted from *Utah Press*; [Fred Bulkley] to Thomas F. Daly, Leadville, June 18, 1898, John F. Campion Papers, Western History Collection, University of Colorado Libraries, Boulder (hereafter cited as CU).

[8] Diary of M. O'Gorman, March 10 and March 17, 1901, Special Collections Department, University of Arizona Library, Tucson (hereafter cited as UA); R. Orchard Old to Subcommission on Mining of the U.S. Industrial Commission, Georgetown, April 22, 1899, Robert Orchard Old Papers, SHSC.

[9] Evadene Burns Swanson, "Where's Manhattan?" *Colorado Magazine* 48 (Spring, 1971), 146–158; Patrick C. Henderson, "The Bradshaw Bonanza," *New Mexico Historical Review* 38 (April, 1963), 151–162.

storehouses, bunkhouses, offices for management personnel, telephone and telegraph extensions, electrical generators, smelters, concentrators, and stamp mills. They were in fact vast industrial establishments that applied improved technology to the removal and reduction of the ore, and thus merited the name "extractories."[10]

One important distinction between small and large mines was the amount of mechanization employed. The small ones utilized manual labor. The mine was a single shaft or, if practical, a tunnel to the ore vein. The technology was rudimentary; men working in small mines drilled their own holes, loaded them with powder, fired the shots, shoveled or "mucked" the ore into buckets or cars, and raised it or pushed ("trammed") it to the surface opening (a simple sketch of this type of mine can be found in Figure 1). Small mines were labor intensive and, lacking occupational specialization, provided a varied work experience of which windlassing, or hauling up the ore buckets, was the most onerous task. George Parsons, who did it all at one time or another, described windlassing as "too dreary and monotonous."[11]

The large mines gradually came to employ the vast majority of the miners. As early as 1860 when mineowners introduced steam-powered hoists in the Comstock mines, mechanization began to alter mining techniques. Subsequently, cages replaced ore buckets; wire cable, rope; dynamite, gunpowder and nitroglycerin; machine drills, hammers and hand drills; and steam and electricity, human and animal labor. In addition, a host of technological innovations altered the processes of ore extraction. Without these transformations and modifications, mines could not have developed as rapidly, descended as deeply, or employed as many miners as they did, but the exploitation and extraction of the resources might have been more systematic and thorough. Technology was not an unmixed blessing, and it created new problems and dangers for the men who worked underground.[12]

[10] Gene M. Gressley, ed., *Bostonians and Bullion: The Journal of Robert Livermore, 1892–1915*, p. 188; Fred G. Bulkley to Edward Doerr, Leadville, December 29, 1898, Campion Papers, CU.

[11] Parsons, *Journal*, March 15–July 7, 1880, pp. 100–136.

[12] Lord, *Comstock Miners*, pp. 220–221, 227, 336, 365–367, 391–392; Rodman W. Paul, *Mining Frontiers of the Far West, 1848–1880*, especially pp. 67–68; Clark C. Spence, *Mining Engineers and the American West: The Lace-Boot Brigade, 1849–1933*, pp. 385–388; Young, *Western Mining*, pp. 151–177,

Inasmuch as most miners found employment in the larger mines, a focus upon the environs, the work, and the attendant dangers in the extractories (sketched in Figure 2) gives the best picture of the typical miner's situation. Before actually going underground, miners might change into special working clothes. Some mines provided special change rooms fitted with lockers and showers for this purpose. As gambits of employers interested in eliminating theft, change facilities were most common in precious-mineral operations. Deep mines like those on the Comstock provided these accommodations because men had to wear special garb to work in the hydrothermal depths. At other mines the pressure for change rooms came from the miners themselves, while the need for them was denigrated by parsimonious management.[13]

When dressed for work, the miners passed through the mine portal or shaft hole (the point of demarcation between topmen and miners), carrying with them their dinner pails, their allotment of candles (usually three), a box of matches, and a candle holder or lantern for use underground. Candles remained the primary light source well into the twentieth century, and as one miner remembered, they "flickered every time one moved." Though miners could walk to their appointed places in the mines, transportation was often provided. Not only were ore cars and giraffes faster, but, since they were in the tunnels anyway, they in fact posed hazards to anyone who walked rather than rode in them. One visitor who hiked into Gilpin County's Bobtail tunnel described the tedious work of "threading . . .

204–217; Wright, *Big Bonanza*, pp. 89–91, 121–122, 220–249, 385–388; Mark Wyman, "Industrial Revolution in the West: Hard-Rock Miners and the New Technology," *Western Historical Quarterly* 5 (January, 1974), 39–45; Walker De Marquis Wyman, "The Underground Miner, 1860–1910: Labor and Industrial Change in the Northern Rockies," (Ph.D. dissertation, University of Washington, 1971), pp. 15–16, 378; Irving Hale, "Electric Mining in the Rocky Mountain Region," *Transactions of the American Institute of Mining Engineers*, reprint, Mining File, WHRC.

13 Wright, *Big Bonanza*, pp. 225–226; interview with Albert J. Frantz, Denver, April 6, 1972. Eben Olcott described the Comstock's special apparel as consisting "of a loose pair of covalur trousers and a loose flannel blouse, woolen socks and rubber boots with a rubber coat to put on as returning to ordinary temperatures." (Olcott to Pheme, Virginia City, September 9, 1881, Eben E. Olcott Papers, WHRC).

along this dark cavernous depth with no light but a bit of a wax candle that I had to shield with my hands against the current of air steadily flowing outwards."[14]

Descent into the main shaft was accomplished by means of either ore buckets or cages, which were open platforms that could accommodate twelve men or an ore car. Some mines provided ladders, but cages or buckets operating in the same shafts made climbing unnecessarily dangerous except in time of emergency. The frequence of travel down and up the shaft in a bucket or on a cage did not prevent it from being one of the most dramatic events associated with mining. The buckets and cages were attached by rope or, later, wire cable to powerful hoisting machinery. Communication between the men in the depths and the operators or hoisting engineers was carried on by prearranged bell signals. The bell was positioned near the hoist, and wire and rope connected the bell to the levels of the mine. In the twentieth century, electricity was employed to transmit signals to the surface.[15]

Riding underground in a bucket or on its rim was dangerous and led one miner to conclude that "three hundred feet or so of that, when the hoist was working jerkily with an occasional back slip, engraves itself on one's memory." As a mine developed, cages (some had three separate platforms) replaced the bucket, because they made more efficient use of mine cars and moved more men. Once aboard the cage, miners prepared to plunge down. As the warning bell sounded, the cage dropped into the dark shaft. The only light came from lanterns affixed to the cage itself and from momentary windows of light that emanated from the stations passed on the way down. Likened by some miners to being buried alive, the fall produced only muted sounds, the smell of damp ground, and the rush of air; then from the

[14] Frank A. Crampton, *Deep Enough: A Working Stiff in the Western Mine Camps*, p. 42; Diary of I. J. Baldwin, October, 1873, SHSC. Giraffes are specially designed ore cars with the rear of the cargo bin elevated so that the cars can operate on inclines without cargo loss. The elevation is attained either with lift rods, on an adjustable bin, or with oversized rear wheels.

[15] Wright, *Big Bonanza*, pp. 222–224; Lord, *Comstock Miners*, pp. 221–222; Mrs. Hugh Brown, *Lady in Boomtown: Miners and Manners on the Nevada Frontier*, pp. 48–49; Keena, "Cripple Creek," pp. 3–4, Keena Papers, SHSC; Moudy, "Cripple Creek Miner," p. 380.

pit of the stomach came the sinking feeling that accompanied the rapid fall. "Our first thought is that between us and the bottom of the shaft—fifteen hundred feet below—we have nothing but the frail platform of the cage, and instinctively we tighten our grip upon the iron bars of the cage, determined that, should the bottom drop out, we will be found hanging to the upper works of our strange vehicle."[16]

The dim light of the lanterns encouraged optical illusions of a moving shaft and timbers that seemed to "be darting swiftly upward." At great depth the cable showed a buoyant elasticity as it was stretched taut over hundreds of feet. These phenomena, coupled with those noted above, provided passengers with an eerie sense of timeless unreality and impending doom. Even the seasoned miner breathed a sigh of relief when at last the cage reached his level.

The station was an enlarged portion of the shaft where the shaft and level met. Here miners disembarked or later embarked for their journey to the surface; here too, the ore cars were loaded onto and unloaded from the cage. Lighted by kerosene lanterns, the station served as a repository for mining supplies, water barrels, and any excess clothing the miners carried with them. Here men chatted briefly and relaxed while awaiting the cage.[17]

Leaving the station, the miners moved on to their respective places of work. Usually afoot, they followed the drifts and crosscuts to the vein or to the point at which the last shift had stopped work. Their job underground was to locate or reach the vein and then to extract the ore from it. Dead work, which included shoring up the underground works and removing accumulated waste rock, and exploratory work were essential components of the mining operation. Actual ore removal was always intermittent, since miners also had to facilitate the removal, search for extensions of the vein or new veins, and deepen or widen the shaft. This was work that did not immediately yield ore, but it was done in order to expedite further ore removal. Thus men were stoping, drifting, sinking, crosscutting, mucking, tramming, timbering, and hoisting; each man engaged in his assigned task. (A satirical categorization of the types of workers can be found in Appendix A.) The ore moved through this underground

---

[16] Wright, *Big Bonanza*, pp. 222–228.
[17] Ibid., pp. 227–228, 231; Lord, *Comstock Miners*, pp. 212–213.

circulatory system until it reached the surface, there to continue on its way to mills or smelters.[18]

Climatic conditions in mines varied considerably. Deep mines often extended below the water table and were damp and humid. Water seepage proved a constant irritant to the workers, and large quantities of water could virtually eliminate access to underground regions. Underground temperatures were constant, thus being cooler than the outside air in summer and warmer in winter. No wider and no higher than necessary, the underground passageways were coarsely constructed, and sharp wood splinters or rocks protruded into the corridors. The floors were littered with bits of rock, wood, human and animal offal, and miscellaneous debris. The smell of powder smoke and decaying lumber was also present. These conditions were further exacerbated by poor ventilation, a concomitant of hasty or careless development. The best miners often refused to work in improperly ventilated works, where accumulated smoke and dust irritated eyes and noses and unvented gases threatened life.[19]

Deep in the hydrothermal regions the underground environment was noisome. In the 1880's the Comstock mines reached depths of half a mile or more and recorded temperatures of 107°F. Under those conditions miners stripped to breechcloths, thick-soled shoes, and hats to protect the head and eyes. Eliot Lord reported that in 1873 Crown Point miners had encountered water "so hot that eggs could readily be cooked in it." In these ill-ventilated and dehydrating depths, miners drank water and chewed ice. Here, two crews of men worked on each shift; one group labored at the face while the other rested. They experienced noticeable physical discomfort and craved salty foods. At the other extreme were high-altitude mines like Colorado's Stevens Mine where miners lighted fires in the evening so that the rock

[18] Parsons, *Journal*, May 26, 1880, p. 126; Lord, *Comstock Miners*, p. 312; Wright, *Big Bonanza*, pp. 234–235; "Among American Miners," *Chambers Journal*, ser. 5, 8 (September 5, 1891), 572.

[19] Eben E. Olcott to Katherine, Pioche, Nevada, June 1, 1892, Olcott Papers, WHRC; Diary of Cecil C. Morgan, January 22 and 23, and February 11 and 12, 1878, DPL; Diary of Arthur Bowen, October 8, 1882, DPL; Joe Manwaring to E. D. Gardner, Utah, May 31, 1912, and Gardner to Manwaring, June 5, 1912, E. D. Gardner Papers, WHRC; Ernest LeNeve Foster to Jefferson Reynolds, Central City, October 17, 1912, Ernest LeNeve Foster Papers, DPL.

would thaw by morning and thus permit the next day's drilling. Such mines recorded temperatures at or below freezing within the works themselves.[20]

While actual mining began with drilling and blasting, these were rarely the first tasks assigned to an inexperienced worker. Tramming, pushing the ore cars, required the least skill. Mucking or shoveling was the next step toward a career as a miner; since it required an ability to differentiate between ore and waste rock, it assumed some experience underground. If the jobs were combined, with one individual mucking and tramming his own cars, the mucker-trammer was expected to load and move about two cars per hour. If the tasks were separated, the production schedule doubled.[21]

In the narrowest sense the term *miner* applied to those men working at the face and engaged in drilling, loading, and detonating the holes. Drilling was the first step in the complicated procedure of breaking the ore from the face. Although machine drills appeared as early as the 1870's, hand drilling persisted well into the twentieth century. As in the drilling contests, hand drilling took one of two forms, single-jacking or double-jacking, and employed the working versions of the same equipment, hammers and drills.[22]

Both single- and double-jacking had their peculiar advantages. Single-jacking was ideally suited to narrow stopes, where there was little clearance and the drilling was usually overhead. But since it required constant effort by one miner, it was the ultimate in hard work, or as one miner remarked, not a job "for softball players." Double-jacking was less tiring since both men took their turns at the hammer, and it also allowed each man to maximize his effort, whether swinging the hammer or turning the drill. As in the contests, careful preparation and examination of the rock paid handsome dividends to the skilled miner, who pointed his holes to get the maximum effect. Lacking his

---

[20] Eben E. Olcott to Pheme, Virginia City, September 9, 1881, Olcott Papers, WHRC; Wright, *Big Bonanza*, pp. 241–242, 248; Lord, *Comstock Miners*, pp. 212, 389–391, 393–396, 400–401; R. Weiser, "Permanent Ice in a Mine in the Rocky Mountains," *Engineering and Mining Journal* 17 (1874), 358.

[21] Emory Smith, "Forty Years of Progress in Mine Safety," State of California, Department of Industrial Relations, Division of Industrial Safety, *Bulletin* (1963), pp. 3–4; Crampton, *Deep Enough*, pp. 11, 42–43.

[22] Parsons, *Journal*, March 20, 1880, p. 105.

skill, the novice had to compensate with physical exertion, frustration, and premature fatigue. One tenderfoot regarded drilling as "a trial I'll venture to say—that very—very few of my condition in life have ever experienced."[23]

With drilling completed, the miners next turned to loading and firing the powder. As in drilling, loading and detonating the charges required concentration, skill, and a heavy dose of caution. For, as George Parsons so aptly put it: "This again is dangerous work. We use Giant Powder. Has to be handled carefully. Tamping is done very nicely—not too much—or cap explodes and one is likely to go flying out of the shaft. Black powder is not so dangerous but plenty bad enough." Frank Crampton remembered that these skills were "taught as an art." Before the sticks of dynamite or black powder were inserted, the casing material was slit several times from end to end. Thus as the powder was pushed forward, it would flow out and fill the hole. To be effective, the powder had to be tightly packed into the hole. The type and quantity of explosive used depended on both the geology of the mine and the result desired. With the powder snugly tamped, the detonating device, called a cap or primer, was inserted. Given its nature and purpose, the cap was volatile; special care accompanied its placement on the powder. Wooden sticks were the recommended tamping tools, because wood could not strike a spark and thereby cause premature detonation. Finally, more powder was packed around and over the cap, and the entire charge was covered with mud.[24]

Detonation followed the tamping and priming. It was done at the end of the shift to allow the dust and gases to settle or dissipate before the next group of miners entered to muck the ore and set another round of explosives. The off-going men kept a careful count of the explosions so that they could warn their replacements if one of the charges had failed. It was serious business, as Frank Crampton remembered:

Almost as important as loading was the way a round should be spit, "light the fuse" in tenderfoot vernacular, and fired. I quickly learned to cut the spitting end of a fuse wide, and deep, and to shake the fuse pow-

[23] Crampton, *Deep Enough*, pp. 42–44; Parsons, *Journal*, March 18, March 31, and November 3, 1880, pp. 104–105, 109, 179.
[24] Parsons, *Journal*, March 18, 1880, p. 105; Crampton, *Deep Enough*, pp. 44–45.

der into the V of the cut. The fuse spit faster when that was done right, because there was more powder to take fire, and it was seldom necessary to waste time making a new cut and try spitting again; one could get away sooner from a face that was ready to go.

Spitting with candles was something. The backfire of a spit fuse would blow out the candle, and it took time to light it again, all the while the other fuses that had been spit were still burning and getting nearer to letting all hell loose. Even snuffs, those two-inch or less pieces of burned candles that were lighted and set close by, didn't help much, for it took time to reach for one to go on spitting. The danger of heart failure ended when I learned to make notches in a short fuse and when fire came from the notch, spit with it; it was fast and sure, and I could walk away from a spit round, not run.

These shift-ending explosions attracted the attention of all persons underground. Tonopah's Mrs. Hugh Brown remembered them as "a sudden sensation, a feeling that our bodies were being pressed in from all sides at once." She noted that "an atmosphere of unmistakable tension filled the darkness." This was always serious business; the miners were relieved when it was over and they were safe and sound, with their shift at an end.[25]

Tramming, mucking, drilling, and blasting did not exhaust the occupational demands made upon these men. Like other craftsmen, they had to keep their equipment in optimum condition. In small mines they sharpened their own drills and refurbished other tools. Large mines, on the other hand, had blacksmiths and tool tenders who handled these chores. On occasion the miners did their own timbering; when an individual accepted a contract that paid a specified rate for every foot advanced, timbering was usually obligatory.[26]

One particularly dangerous task was the removal of obstructions from ore chutes or mill holes. These chutes connected two levels in large mines. They facilitated ore removal by providing a tunnel linking the stopes of one level with the ore cars of the level below. Frequently large chunks of ore and waste rock lodged in these passageways, blocking the flow of ore. Opening them entailed climbing up to the

[25] Crampton, *Deep Enough*, p. 45; Brown, *Lady in Boomtown*, p. 49.

[26] Parsons, *Journal*, November 1, 1880, p. 178; Diary of Bowen, October 1, 3, 9, 11, and 13, 1882, DPL; Diary of O'Gorman, October 5 and October 7, 1901, UA.

bottleneck from the lower level, setting a dynamite charge, and then crawling back down the chute before the explosion occurred. This activity was extraordinarily hazardous; the rocks might shift and tumble upon the unfortunate miner, his fuse might burn too rapidly, or he might have difficulty getting back down the chute. Even the double pay often used as incentive was not sufficient motivation for most men; here was a task to be avoided.[27]

From time to time miners were assigned other duties, such as replenishing underground supplies, grading the surface around the shafts, windlassing material into the shaft, raising ore, constructing surface buildings, sawing lumber, and working on prospect holes. If weather was inclement topside or the tasks unpleasant, the miners quit work to prospect or to finish personal projects.[28]

Working at the varied tasks of their trade, miners were a proud group. Eliot Lord likened them to "models for a sculptor. . . . Their variety of motions made them a troop of athletes." They were usually quiet, concentrating their attention and strength on the tasks before them. The working conditions and problems miners faced were lamented by a mining bard, B. A. Price, in a poem entitled "The Driller's Troubles."

> The ground is as hard as the devil,
> The steel won't cut at all,
> The smith knows a lot about temper!
> The boss ought to give him a call.

> [*sic*]
> The next one's as soft as lead
> If he don't give us stuff we can work with
> I'll bend a length over his head.

> My partner blocked up like a farmer,
> The bar's coming down by the feel,
> The air hose and water is brakin',
> And I'm most as wet as a seal.

[27] Crampton, *Deep Enough*, pp. 85–87; Robert Livermore, "An Autobiography," p. 128, Robert Livermore Papers, WHRC.

[28] Diary of O'Gorman, October 2, 5, 7, and 8, and November 29, 1901; UA; Diary of Albert B. Sanford, June 24, August 8, and August 10, 1881,

The machine takes oil like a furnace,
   And it blows it all over the face;
Some son of a gun has swiped my wrench
   And left a bum one in its place.

We'll ne'er get the round in by tally,
   It's a cinch that the shifter is sore.
If the muckers had gotten the dirt out,
   We'd have finished it long before.

The powder and caps are rotten,
   The fuse is a genuine fake,
The air is chock full of water,
   And the blasted ground won't break.

There's no need in sweating your life out,
   This job is awfully tough,
So I think it's time to taper off,
   I guess she's in deep enough.[29]

"The Driller's Troubles" lies somewhere between doggerel and poetry, but its author conveys a sense of the driller's pride at being a miner as well as sketches a caricature of the frustrations he faced. Like thousands of other hard-rock miners, the poetic figure would probably have trudged to his work at the mine, grumbling about another long shift. Sometimes the daily routine of descending and mining became monotonous, but, as previously explained, miners like George Whitwell Parsons believed themselves superior because they could do it, and contemporary observers like Eliot Lord praised miners' courage and accomplishments underground. The strenuous activities of drilling, blasting, and tramming in the man-made caverns spawned a kind of arrogance, tempered always by the dread of impending disaster.

Albert Byron Sanford Papers, SHSC; Foreman Hulburn's Report, Eagle River Consolidated Mining Company Papers, SHSC.

[29] Lord, *Comstock Miners*, pp. 90–91, 311–313; B. A. Price, "The Driller's Troubles," *Miners' Magazine* 19 (July, 1918), 6, reprinted from *Mining and Scientific Press*.

5.

The Deadly Depths

But the arms that deliver the sturdy stroke,
Ere the shift is done, may be crushed or broke,
Or the life may succumb to the gas and smoke,
Which the underground caverns fill.[1]

BEYOND the personal frustrations that accompanied mining loomed the occupational hazards. From the moment that a man left his home in the morning, he encountered risks linked to the peculiar environment of the mining towns and to the special perils of mining. Like a shadow, danger followed him wherever he went and in whatever he did. Even trudging to work with his lantern in hand, a miner risked injury in districts honeycombed with old shaft holes, abandoned tunnels, and other remnants of the initial mining excitement. Then, in this mountainous region, there was the possibility of slipping or sliding down a steep slope. J. J. Jones, a Leadville man, suffered such a fall and lost thirteen days of work. The perils of western life plagued other miners, some of whom sustained injuries when thrown from horses or wagons. For example, wagon accidents crippled Henry Mayfield of Leadville and killed Mat Ragan of Manhattan, Colorado. Although commuting was their least dangerous work-related activity, 3 percent of the Ibex Mine laborers (8 miners in an average work force of 268) suffered accidents connected with the trip to and from work in 1899–1900.[2]

[1] Alfred Castner King, "The Miner," *Mountain Idylls and Other Poems*, p. 46.

[2] Kenneth Fahnestock to Thomas Daly, Leadville, April 7, 1900, November 11, 1901, and January 2, 1902, John F. Campion Papers, Western History Collection, University of Colorado Libraries, Boulder (hereafter cited as CU); Evadene Burns Swanson, "Where's Manhattan?" *Colorado Magazine* 48 (Spring, 1971), 156; "Ibex Accident Report for 1899," October 31, 1900, Campion Papers, CU.

The most common source of accidents among miners was falling objects. Working underground and often in haste, miners were often injured by falling rocks, timbers, equipment, and other objects. Underground chambers always contained loose rock; it required merely the accidental convergence of a miner and a slipping rock to produce a tragedy. For example, Albert Byron Sanford recorded that a comrade and fellow miner named Broad was struck twice in eight days by falling rock. The first incident caused only slight damage, but the second rock gashed his head and "hurt him badly."[3]

The miners' limbs were most vulnerable to injury; the head and the torso, less susceptible. Constantly scraped, scratched, or crushed, fingers and toes were bruised and, not infrequently, pulverized by falling rocks and timbers. One Arizona mucker lost seventeen days of work when he was struck on the toe by a rock that plummeted from the ceiling. In Leadville, George H. Stewart suffered serious injury to the muscle tissue of his left arm when a falling rock struck him; the accident report concluded that he would probably be incapacitated for six weeks. In 1894 one commentator noted, "The leading source of accident in all kinds of mining is a prosaic but deadly one—the fall of rock from the roof and sides of workings, falls of ore in the stopes of metal mines."[4]

Accidents due to falling objects were generally attributed to employee negligence or to unavoidable occupational hazards. (Whether this was always the case will be discussed at greater length in chapter 7.) Miners were especially fearful when crossing open shafts or working in areas where the roof was far above the work area because objects there fell great distances. Even the smallest object falling down a deep shaft became a potential source of tragedy. In one instance, a rat reportedly fell two thousand feet and struck an unsuspecting miner, killing the man instantly. When visitors or careless surface

[3] Diary of Albert B. Sanford, August 14 and August 22, 1881, Albert Byron Sanford Papers, State Historical Society of Colorado, Denver (hereafter cited as SHSC).

[4] "Report of Serious Accidents," November, 1917, February, 1918, and July, 1918, Arizona Copper Company Ltd. Papers, Special Collections Department, University of Arizona Library, Tucson (hereafter cited as UA); State Accident Reports filed by W. S. Crocket, Ibex timekeeper, Leadville, November 2, 1908, and October 3, 1911, Campion Papers, CU; Albert Williams, Jr., "The Miner and His Perils," *The Chautauquan* 18 (January, 1894), 429.

personnel dislodged small rocks or dropped tools, natural acceleration made these items as deadly as projectiles fired from a gun. Rocks killed Joseph Oates on the Comstock and James Hoskins of Central City, Colorado. Such disasters prompted mining companies to caution visitors about the careless handling of rocks; individuals were kept from the shafts and reminded to handle equipment with care. Sometimes such accidents contributed only indirectly to a death. In 1899 Mrs. Margie Alley lost her husband after a rock hit him in the neck; he died, however, not because of the injury but because he contracted smallpox in the hospital while receiving treatment. Even falling dust particles proved dangerous, as mucker Rosendo Edesa discovered. When Edesa looked up just as some dust trickled from the ceiling, it lodged in his left eye and damaged the cornea, costing him twenty-two days of work. Falling objects accounted for nearly 9 percent of the fatalities on the Comstock and about 20 percent of the reported injuries. In the mines of the Arizona Copper Company, Ltd., falling objects combined with caving to produce more than 50 percent of all injuries and fatalities (see appendix B). Statistics on mining accidents compiled by the U.S. Bureau of Mines over the nine-year period between 1911 and 1919 show the frequence of various causes of accidents regionwide (see appendix B).[5]

Not only falling objects but also their own falls threatened miners. A falling rock and a slip of the foot once combined to injure Fred Balco of Leadville. While working at the top of a raise (a steeply inclined passageway connecting one work area with another on a higher level), he was struck by a rock that caused him to lose his balance and fall through the hole at the top of the raise. Though seriously injured, he was more fortunate than Celeste Marchetti, an Arizona miner who was struck by falling timber while he was standing on a platform in a chute. Knocked off the scaffolding, he fell five hundred feet to his death.[6]

[5] Williams, "Miner and Perils," p. 430; Arthur Cecil Todd, *The Cornish Miner in America*, pp. 162, 192; Fahnestock to Daly, Leadville, December 5, 1899, and J. Campion to Mrs. Margie S. Alley, September 6, 1899, Campion Papers, CU; "Report of Serious Accidents, February 1918," Arizona Copper Company Papers, UA.

[6] Fahnestock to Daly, Leadville, May 26, 1899, Campion Papers, CU; "Report of the Mine Inspector for Six Months Ending September 30, 1915," Arizona Copper Company Papers, UA.

Most dangerous were falls in shafts, ore chutes, winzes, and raises, where miners customarily used mine timbering (or lagging) as bridges. These boards were rarely anchored, and the slightest slip sent the miner plunging into the depths. George Parsons recorded the fate of one lucky individual who slipped from a board into a hole but escaped serious injury because the hole was shallow—only ten feet deep. Thomas Stivens, a Georgetown miner, fell to his death when he unaccountably toppled from a firm plank he had set across the main shaft of the Pelican Mine. Some men escaped from such falls with their lives, as did Cornishman Richard Pearce, who nonetheless spent two years in a hospital recovering from a one-hundred-foot fall that dislocated both ankles, broke one leg at the hip, and tore leg muscles. Albert Williams, Jr., concluded that experienced miners were more likely than newer hands to be injured in accidents caused by carelessness because, familiar with their surroundings and the work, the seasoned workers became negligent.[7]

Drilling and loading holes and handling explosives posed other hazards. Throughout this period the mining journals, newspapers, state bureaus of mines or labor, and eventually the U.S. Bureau of Mines devoted particular attention to these dangers. Miners were cautioned against using metal implements in handling and loading charges, against careless placement and storage of powder, and against complacency. Yet mishaps persisted; ignorance, carelessness, and short cuts combined with circumstance to produce human tragedy and suffering.[8]

Ignorance and negligence contributed especially to accidents stemming from the handling of powder. For example, Tombstone's Sunset Mine was the scene of an explosion in 1880 when dynamite

[7] George Whitwell Parsons, *The Private Journal of George Whitwell Parsons*, May 26, 1880, p. 126; "Fatal Accident," *Georgetown Colorado Miner*, August 22, 1874; Todd, *Cornish Miner*, pp. 164–165, 192, 220–222, 261–262; Williams, "Miner and Perils," p. 430.

[8] Albert H. Fay, comp., "Metal-Mine Accidents in the United States during the Calendar Year 1915," U.S. Bureau of Mines, *Technical Papers* 168 (1916), pp. 85–86; Clarence Hall, "Permissible Explosives Tested Prior to January 1, 1912, and Precautions to Be Taken in Their Use," U.S. Bureau of Mines, *Miners' Circular* 6 (1912). These works specify the basic rules for the safe handling and care of explosives.

stored in the blacksmith shop was detonated by an errant spark that touched off the blasting caps. Eventually, disasters of this kind made management more conscious of the need for care in the issuance and handling of powder. A. J. Anderson described the prevailing practices in the Socorro Mining and Milling Company of New Mexico: "We have not made primers underground for at least eighteen months, and the rule of never keeping caps and powder close to each other anywhere is rigidly enforced." As management personnel observed elsewhere, "the best insurance against accidents is care in operation and responsible protection against the possibilities of them."[9]

Drilling holes and loading shots produced an even greater assortment of injuries. One common cause of accidents was the tamping of powder into the hole with a metal cleaning spoon. Michael Sullivan, a friend of Frank Crampton, was killed when his metal spoon apparently struck a spark that ignited the charge. One expert attributed many so-called premature explosions to the widely held assumption that one foot of fuse burned for one minute, an error of more than 25 percent according to later investigations, which fixed the actual rate at a foot every forty-two or forty-three seconds.[10]

Overloading the charge was another source of danger. Henry E. Wood remembered overloading one hole and, fortunately, only being showered with dirt, rock, and debris. Such overloading sometimes had more drastic consequences. Nothing, however, was as deadly as misfired holes. Miners were admonished to count carefully all the charges as they detonated. If twelve charges had been set, it was imperative that all twelve be accounted for. A failed charge meant danger no matter how it was handled. If the blasters returned to the working face to find the powder and cap, they ran the risk of being injured in a tardy explosion since fuse was not always of uniform quality and might simply be burning more slowly than usual. If, like Criso Gal-

---

[9] Parsons, *Journal*, August 14, 1880, p. 152; A. J. Anderson to W. Rowland Cox, June 19, 1913, and J. E. Hulshizer to D. B. Scott, August 8, 1918, Blumenthal Papers, Special Collections Department, Zimmerman Library of University of New Mexico, Albuquerque (hereafter cited as UNM).

[10] Frank A. Crampton, *Deep Enough: A Working Stiff in the Western Mine Camps*, p. 234; Emory Smith, "Forty Years of Progress in Mine Safety," State of California, Department of Industrial Relations, Division of Industrial Safety, *Bulletin* (1963), p. 4.

lardo, they attempted to "reload a hot hole," they might also be killed. On the other hand, if they ignored or counted inaccurately the number of blasts, the men on the next shift were exposed to unnecessary danger. Digging with pick and shovel in poorly lighted underground works, a man could easily pick into a charge before he realized its existence. Or an unsuspecting miner might drill into the missed hole. Robert Livermore long remembered the gruesome scene that met his eyes when he stumbled upon the place where two machine miners had been killed as they drilled into a missed hole.[11]

Drilling itself, whether by hand or by machine, could also cause personal injury. Bits of rock or steel and dust broke during the drilling process; frequently, the fragments snapped back, striking the miners in the body or the face. Writing about his experiences in 1904, Livermore noted: "[A] piece of steel flew from the drill I was using and lodged in the exact centre of my eye. Fortunately, it didn't pierce the cornea, or I would have been minus one eye, but it gave me trouble enough. Continuing to fester and to blind me, it at last forced me to go to Denver, where a remaining fragment was extracted by magnet." Picks and drills sent flying percussion chips that caused serious eye injuries. In 1902 Kenneth L. Fahnestock wrote to insurance adjuster Thomas F. Daly "that a large percentage of the minor accidents in our mine are from this same cause, and while there may seem to be a great many of them I can see no way to avoid it."[12] A later generation found safety goggles the best solution.

While battered hands and broken arms were the trademarks of hand drilling, machine drilling produced its own distinctive injuries: strained muscles, especially backs, and bruised arms and fingers. Serious accidents occurred when the heavy percussion drills slipped while in operation. Furthermore, drilling and blasting produced dust: dust that the miners breathed, dust that irritated the lung tissue, producing increased susceptibility to pulmonary diseases and

[11] Henry E. Wood, "I Remember," p. 25, Henry Ellsworth Wood Papers, Huntington Library, San Marino, California; "Report of Mine Inspector for Six Months Ending March 31, 1914," Arizona Copper Company Papers, UA; Robert Livermore, "An Autobiography," p. 128, Robert Livermore Papers, Western History Research Center, University of Wyoming, Laramie (hereafter cited as WHRC).

[12] Livermore, "Autobiography," p. 82, Livermore Papers, WHRC; Fahnestock to Daly, Leadville, August 13, 1902, Campion Papers, CU.

leading ultimately to miners' consumption. Eventually, lung disease would be recognized as the miners' most serious affliction.[13]

Techniques of mining and the instruments of extraction changed so drastically in the years between 1860 and 1920 that the industry itself was transformed. Yet, in another sense, the basic procedures changed very little; machines had aided or displaced men rather than transformed the process itself. Mining underwent continuous industrialization. Dynamite, air drills, electricity, hoisting cages, underground railroads, drainage tunnels, square-set timbering, block caving techniques, and finally open-pit (steam-shovel) mining had made possible mining at greater depths, mining of lower-grade ore, mining of heretofore unreachable ore bodies. But often innovation was accomplished at great cost; Leo J. Keena estimated that the mines of Cripple Creek were spending about five dollars to extract one dollar's worth of ore. For the men underground, technological change exacted more terrible costs: lives lost and injuries sustained.[14]

Walker D. Wyman, Jr., believes that these costs have been largely ignored, and in one important sense he is correct. Insufficient attention has been given the laboring men and the sacrifices they made in the expansion and development of the western mineral industry. As will be seen, the new technology did pose new hazards for them. However, in a broader sense an emphasis on problems of the new techniques misses the central points. Without the changes, the men would not have found jobs. Even more importantly, rather than simply tallying the number of injuries caused by the new methods, one should compare, if possible, preindustrial statistics with those of the industrial period. Were cages, air drills, dynamite, deeper mines, electrical power, square sets, steam shovels, and block caving more hazardous than the mining techniques and implements they displaced? For example, were cages more dangerous than ladders and buckets that dangled at the end of horse-powered whims? A full answer to these questions would require extensive research into the newspapers and mining periodicals of the time. For a tentative answer, however, the accident

[13] Daniel Harrington and A. J. Lanza, "Miners' Consumption in the Mines of Butte, Montana: Preliminary Report of an Investigation Made in the Years 1916–1919," U.S. Bureau of Mines, *Technical Papers* 260 (1921), 17–19.

[14] Leo J. Keena, "Cripple Creek in 1900," p. 2, in Leo J. Keena Papers, SHSC.

statistics of the U.S. Bureau of Mines offer valuable indices of the dangers in mining in the years after 1910. Insofar as these statistics permit comparison between technologically advanced and more primitive mines, the statistics indicate a much higher fatality rate in the smaller, less sophisticated operations. In 1912 the adjusted fatality rate for the small mines was 7.05 men killed per 1,000 men working a 300-day year; the rate in the large mines was 4.02 per 1,000. This suggests that the dangers of technology were no greater and may have been less than those of more primitive mining.[15]

The evidence suggests that, at first, each new development posed extraordinary hazards, stemming from both the ignorance of the miners and the inadequate procedures for evaluating the performance of the new machines, methods, and materials. However, as dangers manifested themselves in injuries and fatalities, changes were made in the equipment and safer operating procedures were established. Nowhere is this better illustrated than in the development of the so-called safety cage.

Cage and bucket accidents were common into the 1920's. The U.S. Bureau of Mines' metal-mine accident statistics for the years 1911–1919 indicate that cages were the most frequent source of identifiable shaft-related accidents, consistently exceeding falls into the shafts, the next most serious category. However, the number of cage accidents did decline over the years, as safety improvements were made. Initially, ropes used to hoist buckets or cages posed serious problems. They frayed easily and thus required constant supervision. Where rope linked whims (horse-powered winches used in hoisting) to buckets, the danger persisted, but on cages wire cable quickly replaced rope. By the 1860's wire cable was used exclusively and ended further experimentation with rope. However, cable too deteriorated, especially when used to hoist ore cars or ten to twelve men at a time. As William Wright explained, a broken cable was "the most dreaded of accidents. There was no dodging when a cable parted. All who were on the cage must go to the bottom of the shaft." In the early

---

[15] Walker De Marquis Wyman, Jr., "The Underground Miner, 1860–1910: Labor and Industrial Change in the Northern Rockies" (Ph.D. dissertation, University of Washington, 1971), pp. 1–31, 176–229, 378, 388; Albert H. Fay, comp., "Metal-Mine Accidents in the United States During Calendar Year 1912," U.S. Bureau of Mines, *Technical Papers* 61 (1913), 13.

years, falling cages accounted for nearly 10 percent of all fatalities on the Comstock and spurred development of the safety cage. Available by the 1870's, these cages had either safety clutches or a device that forced "eccentric, toothed wheels" into the wooden sides of the shaft, thus promptly stopping the cage when cables parted. According to Eliot Lord, the Comstock's last fatality attributable to a falling cage occurred in 1866, and only one serious injury marred the next fourteen years. Thereafter, though still dreaded, falls due to broken cable had been virtually eliminated. Subsequent accidents resulted either from negligent maintenance or from cable snags.[16]

The persistence of cage accidents into the twentieth century suggests that the physical construction of the cage itself was not the sole source of cage-related accidents. Cages were operated by hoisting engineers on the surface and by underground workers, called cagers, situated at the various stations. These men assumed responsibility for the lives of their fellow workers; the engineers inspected and replaced damaged cables and regulated the speed of the cages. Their "every movement usually meant the life or death of some man or men." Standardized bell signals directed them; there were signals for hoisting, lowering, stopping, blasting, danger, and so forth. Ordinarily, engineers acknowledged the signal sent from below by jiggling the cage. This system of signaling demanded constant vigilance on the part of the hoist man, because if a signal was confused or ignored, the consequences could be disastrous.[17]

General inattention, negligence, or confusion by hoisting engineers was a common source of accidents. For example, one Colorado miner recalled that failure to check the cable fastenings caused the deaths of five miners when their cage dropped down the shaft unattached to the hoisting reel. Overwinding was another danger, in which a careless engineer hauling men out of the shaft failed to slow and stop the cage as it neared the surface, dooming men to be crushed in the heaves or dashed to the floor of the shaft house. One such acci-

[16] Albert H. Fay, comp., "Metal-Mine Accidents in the United States . . ." for calendar years 1911–1919, and William W. Adams, "Metal-Mine Accidents in the United States during the Calendar Year 1919," U.S. Bureau of Mines, *Technical Papers* 40, 61, 94, 129, 168, 202, 224, 252, and 286 (1913–1921); William Wright, *The Big Bonanza*, pp. 229–230; Williams, "Miner and Perils," pp. 430–431; Lord, *Comstock Miners*, p. 404.

[17] Interview with Albert J. Frantz, Denver, April 6, 1972.

dent in Virginia City killed two men and crippled six; one miner grasped the bell rope at the last second and escaped unharmed. Descending could be just as hazardous, as Leadville miner Pat Kelley discovered. He and two companions were jolted when the engineer misread his indicator gauge and crashed the cage into the tenth-level station at the Ibex. The machinery operated by hoisting engineers and directed by cagers exerted strange fascination for some men. Intoxicated by the power and speed of the equipment, some hoisters became irresponsible, operating their engines and cages "as fast as steam [would] drive [them]." Sober and dependable cagers and hoisting engineers were essential.[18]

A moving cage was a dangerous contraption, "a prolific source of danger to life." Any movement by the cage's occupants or shifting of the load could trigger a tragedy. In the days before enclosed cages, an extended arm, leg, or head would probably be ripped from the body. Sometimes miners were pulled into the shaft and pulverized. During the rapid ascents and descents, miners could contract the bends and literally lose control of their own bodies. Some men who died on cages did so as a result of foolishness or anger, however. In Pevine, Nevada, Henry Mason reported, "There has [sic] been several deaths caused by men punching each other off the cages used in hoisting ore up from the lower levels." [19]

Cages endangered not only their passengers, but also miners and station attendants on the various levels. Cages running in the shafts would strike men who had impulsively looked up or down the shaft. William Truscott of Virginia City was struck while bending over to pick up tools at the shaft's edge. Timberman John Isaac was shoring up timber when hit by a passing cage. Kenneth Fahnestock reported that James A. Sullivan of the Ibex was crushed by a descending cage when he stepped into the shaft compartment instead of under the pumpway, a fatal miscalculation.[20]

[18] Interviews of the Colorado Pioneers (Edward A. Krisher), SHSC; Lord, *Comstock Miners*, pp. 402–403; Report of Operative C. S., June 18, 1897, Campion Papers, CU; interview with Frantz.

[19] Lord, *Comstock Miners*, p. 402; "Dinwiddie Died in Cripple Creek," *Miners' Magazine* 18 (June, 1917), 3; Henry H. Mason to Sister, March 24, 1875, Henry H. Mason Papers, Bancroft Library of University of California, Berkeley.

[20] Todd, *Cornish Miner*, pp. 190, 262; Daniel M. Barringer to John Brock-

Just as cages did, mine cars, giraffes, and electric trains created hazards wherever these devices were in use. In the early years, tramming was done by hand or with the aid of mules. Where access to the mine works was through a tunnel, tramming was done by animal or machine power. In either case, ore-train drivers were tempted by power and speed, just as hoisting engineers were. Trammers and mule skinners often accelerated and risked the life and limbs of anyone so unfortunate as to be in their way. Hand tramming required care, as trammers were wont to slip with their loads. Accustomed to pushing cars onto cages already waiting for them, inattentive trammers occasionally pushed their cars into vacant shafts. Instinctively grabbing hold of their falling cars, the trammers followed them into the shaft and to their deaths. Something as seemingly harmless as stooping to pick up a candle could endanger trammers, as Grant Anderson discovered when he was struck full in the back by a second trammer who did not see him or his car in the darkened passageway.[21]

Motorized cars or trains were even more dangerous than hand cars, as the statistics for the Arizona Copper Company, Ltd., demonstrate. Between 1914 and 1919 cars accounted for 13 percent of the accidents in the company's mines. Jesus Salado, a trammer, was killed when his head was crushed between one of the cars and the stulls or support timbers in a drift. Others caught hands and fingers in the cars, strained muscles attempting to put derailed cars back on the tracks, tumbled from moving cars, or, like Salado, were crushed between the cars and the walls or timbers of the drifts. Operating on inclines, giraffes, like cages, were attached by cables to surface machinery. They too sometimes broke loose from their cables and raced back down the inclines at terrific speeds. One such accident in Virginia City's Yellow Jacket Mine killed one miner and injured four others.[22]

---

man, April 2, 1902, and Barringer to R. A. F. Penrose, Jr., April 2, 1902, Daniel Moreau Barringer Papers, WHRC; Report of the Death of James A. Sullivan, September 7, 1898, Campion Papers, CU.

[21] Robert Livermore, "Fun in Mining," p. 8, Livermore Papers, WHRC; Fahnestock to Daly, Leadville, June 6, 1899, Campion Papers, CU.

[22] "Report of Mine Inspector for Six Months Ending March 31, 1914," and "Report . . . March 31, 1915," "Report of Mine Inspector for the Year Ending September 30, 1919," and monthly reports of serious accidents, No-

A host of other technological innovations endangered miners. James Galloway once injured his hand on a pump rod and later was killed by entangling his sleeve and arm in a blower on the 2,700-foot level of the Comstock's Union Mine. Appearing first in the mines in the 1880's, electricity became an important cause of mining accidents. Unshielded electric lines proved troublesome to early miners; Robert Livermore related the tale of an unwitting trammer whose ignorance of electricity proved his undoing. Told to beware of a high-voltage transmission line, the trammer relayed the warning to his partner, but in doing so, he pointed toward and contacted the line with a metal bar.[23]

Inadequate safety precautions combined with ignorance and carelessness to make electricity one of the more dangerous mining innovations. As risks were identified, however, responsible mining companies acted to protect their employees from hazards. W. Rowland Cox, consulting engineer of New Mexico's Socorro Mining and Milling Company, explained electrical safety to H. E. Wheelock, the general manager. Cox told Wheelock to cut the voltage in the electrical lines through the use of transformers and concluded that ". . . it should not be necessary to instruct the trammers that it is not safe to stand in the cars and use a long bar to punch ore out of the chutes. . . . Please see to it that all workmen are cautioned against doing this in the future." Cox was startled to learn less than three weeks later that the changes had not been implemented, and he begged Wheelock to refer to his earlier letter. An accident linked to the high voltage transmitted underground had spurred Cox to send the safety instructions in the first place.[24]

Unpredictable and catastrophic hazards included bad air, cave-ins, floods, mine fires, and mining diseases. Bad air, which had plagued mining operations since ancient times, resulted from inadequate ventilation, gases accumulated from blasting powder, or the

---

vember, 1917, and January, February, April, and June, 1918, all in Arizona Copper Company Papers, UA; Todd, *Cornish Miner*, p. 201.

[23] John Debo Galloway, *Early Engineering Works Contributory to the Comstock*, p. 21; Livermore, "Fun in Mining," p. 128, Livermore Papers, WHRC.

[24] W. Rowland Cox to H. E. Wheelock, May 12 and June 2, 1917, see also Wheelock to Cox, April 29, 1917, Blumenthal Papers, UNM.

occasional presence of natural gases. Blasting fumes contributed to the death of a White Pine District miner in 1869. When a shot misfired, he descended in a bucket to check on it. Stunned by vapors, he tumbled from the bucket and died from a broken skull. Three decades later another Nevada miner encountered an accumulation of fumes and, recognizing his choking symptoms, climbed a nearby ladder. He too passed out and survived only because he landed on a crossbeam where he was discovered and revived.[25]

Ross B. Moudy, Cripple Creek miner, found mine gas the most bothersome aspect of his work underground. "Many times I have been carried out unconscious and not able to work for two or three days after." Even so, he was fortunate and admitted it; other men died from asphyxiation. Larger mines often developed forced-air ventilation systems to eliminate or reduce the hazard of mine gas. As early as 1876 the officials of the Stevens Mining Company of Georgetown, Colorado, had installed a hand-operated blower to aid underground circulation. Not only did bad air render men unconscious or nauseated, but it also made it more difficult for affected mines to recruit workers.[26] Natural gases were even more troublesome than powder-produced gases. Workers in Central City's Gregory Consolidated Mines encountered an unusual gas in the Black Hawk workings. J. C. Dickey, manager, explained the gas's origin and nature.

The bad air that I spoke of in my letter is a peculiar thing, and is not encountered very often, and sometimes only lasts through small sections of ground. It happens in this way: The men will work in the drift, and the candles [for the detection of gas] will not show it very strong, but as soon as the dirt that they have shot down is stirred at all, there is a gas comes out of it which will do up a man in a good deal more deadly shape than ordinary bad air.[27]

Mine fires and hydrothermal regions also released dangerous vapors. Hazards in their own right, fires also produced carbon mon-

[25] "The Diary of John McQuig," *Nevada Historical Society Quarterly* 6 (Spring, 1963), 22; Uri B. Curtis to Clyde A. Heller, Tonopah, June 18, 1902, Barringer Papers, WHRC.

[26] Ross B. Moudy, "The Story of a Cripple Creek Miner," *Independent* 57 (August 18, 1904), 382; Cyrus Mendenhall to Nat Harris, September 15, 1876, Stevens Mining Company Papers, Western History Department, Denver Public Library.

[27] Dickey to Irving T. Bush, October 16 and October 21, 1897, Gregory Consolidated Mines Papers, CU.

oxide, a dangerous subterranean asphyxiant. Heat and fumes plagued miners on the Comstock and other hydrothermal regions. Poor ventilation, heat, and humidity stifled miners; they gasped for each breath of air and were able to work only brief periods before returning to the cooling stations. As the Comstock miners of the 1870's labored in the heat at these depths, fatalities increased dramatically. Eliot Lord noted that the intense heat killed three miners in a hundred-day period; in 1878 three coworkers died from the heat and fumes in the Gould and Curry Mine. In the same year Thomas Brown was stunned by heat prostration in the Gould and Curry. "He was carried at once to the surface, but continued in a dazed condition for some time, having lost his memory, and babbling like a child."[28]

Pockets of underground water, another source of danger, developed as a result of the downward seepage of rainfall that accumulated in faults and underground reservoirs. Ordinarily more a hindrance than a serious threat, subsurface water deposits necessitated constant pumping and prompted the creation of complex and costly drainage systems like Virginia City's Sutro Tunnels. Leonard Arrington estimates that fully two-thirds of the earnings of Park City's Ontario Mine were spent on pumping. Sometimes the water was more than a nuisance. In 1863 a miner in Virginia City's Ophir Mine picked into a water reservoir, forcing the immediate evacuation of the lower levels. A similar occurrence in 1895 in Central City's Bobtail Mine claimed eleven lives when the men were engulfed before they could escape to a higher level. Water could be a continuing problem, as New Mexico's Socorro Mining and Milling Company discovered. Encountered first in August, 1912, water eventually forced the total suspension of mining at Mogollon's Deep Down Mine in 1915.[29]

Mine fires, like their surface counterparts, terrified miners, their

[28] Lord, *Comstock Miners*, pp. 395, 398–400; Wright, *Big Bonanza*, pp. 374–375, 385–387.

[29] Lord, *Comstock Miners*, pp. 230–243, 297–300, 333–346; Leonard J. Arrington, "Abundance from the Earth: The Beginnings of Commercial Mining in Utah," *Utah Historical Quarterly* 31 (Summer, 1963), 214; report of a disaster in Central City as noted in *Colorado Springs Gazette*, August 30 [1895], collected in Henry F. Lake, Jr., ed., "Remembrances of Pioneer Days in Gunnison County"; W. R. Harris to W. Rowland Cox, Mogollon [N.M.], September 1, 1912, and A. J. Anderson to Cox, Mogollon, June 6, 1915, Blumenthal Papers, UNM.

families, and the community at large. The friends and families of miners below waited anxiously to learn which men had escaped and which had not. The fire-trapped miners themselves rarely had time to consider their plight; the lucky ones made it quickly to safety; the less fortunate usually died almost immediately. Along with its report on metal-mine accidents in 1918, the U.S. Bureau of Mines presented a partial list of metal-mine fires between 1867 and 1919. The five states of Arizona, Colorado, Nevada, New Mexico, and Utah accounted for almost a third of such disasters in the nation, and the list of the afflicted mines and towns included the great names of western mining.[30] Candles, powder, errant sparks, and spontaneous combustion were the most common causes of mine fires.

As in so much else that relates to the mining history of the Rocky Mountain and Great Basin West, Virginia City and Gold Hill were typical in experiencing disastrous mine fires. One 1866 fire began on the 260-foot level of Virginia City's Empire Mine and had soon spread through the works. Smoke and gas forced the evacuation of the Empire and connected workings. Fortunately, this fire and its successor in 1867 caused no serious injuries; however, the conflagration of 1869 was more deadly.[31]

On April 7, 1869, Gold Hill experienced one of the most disastrous metal-mine fires in American history. Indeed the casualties, variously counted at forty-five, thirty-four, and forty-one, make the fire that swept the Yellow Jacket, Crown Point, and Kentuck mines the most serious until the infamous (Butte, Montana) Granite Mountain fire in 1917. A burning candle probably ignited the timbering of the 800-foot level in the Yellow Jacket. As fire consumed the timbering, the roof of the level caved in. This forced the flames, smoke, and gases along the drifts and into the shafts, where the fire spread rapidly. The heat, smoke, and fumes singed and stifled victims. John Murphy, a station man in the Yellow Jacket, was knocked unconscious by the first blast of foul air and smoke. In the Crown Point

[30] Albert H. Fay, comp., "Metal-Mine Accidents in the United States during the Calendar Year 1918," U.S. Bureau of Mines, *Technical Papers* 252 (1920), 72–73. Mine fires were listed for Globe and Jerome, Arizona; Leadville, Silver Cliff, Creede, Black Hawk, and Telluride, Colorado; Gold Hill, Virginia City, Tonopah, and Kimberly, Nevada; White Oaks, New Mexico; and Bingham, Utah.
[31] Williams, "Miner and Perils," p. 432; Lord, *Comstock Miners*, p. 269.

Mine, a rescue cage lowered to the 800-foot level brought up a group of lucky survivors. A second descent netted the three Bickle brothers, none of whom, however, survived. One suffocated; a second passed out in transit and was dismembered as he slipped between the cage and the shaft; the third reached the surface alive but died shortly thereafter.[32]

On the surface the wives, children, and friends of those underground assembled to await word about their loved ones. William Wright described the sorrowful scene at shafts of the three mines. " 'Lost! Lost! Lost!' was the despairing cry constantly uttered by many of the women whose husbands were below." He continued: "As the cage containing the dead bodies rose up at the mouth of the shaft, there was heard a general wail from the women, who could with difficulty be restrained from climbing over the ropes stretched to keep back the crowd. 'Oh God! Who is it this time?' some one among them would be heard to say."[33] In the face of this profound human tragedy, miners and firemen descended into the Yellow Jacket shaft in order to fight the fire and to rescue any injured survivors. Those who were able joined the fire fighters, while others were literally "stunned by the sudden horror of the scene." The mass funeral procession afterward included contingents from the Gold Hill and Virginia City guard units, the Miners' Union, and the Canadian Society, in addition to the bereaved. Eliot Lord later pronounced it the "largest funeral recorded in Nevada."[34]

Mine fires could burn slowly but almost indefinitely in the oxygen-starved atmosphere; the extensive timbering provided an extraordinary amount of fuel. These fires were always dangerous because the introduction of oxygen might at any time respark live coals into another underground blaze. There was the lingering danger from fire-released poisonous gases that reposed in fissures or emanated from decomposing ore and rock. Though apparently not fatal in small

---

[32] The Yellow Jacket, Crown Point, and Kentuck fire of 1869 is discussed in Lord, *Comstock Miners*, pp. 269–277, and Wright, *Big Bonanza*, pp. 126–131. Wright asserts that forty-five died (p. 126), and Lord specifies thirty-four (p. 272). In Fay, "Metal-Mine Accidents, 1918," forty-one casualties are attributed to the 1869 blaze (p. 72).

[33] Wright, *Big Bonanza*, p. 127.

[34] Lord, *Comstock Miners*, p. 271, 273–274.

doses, these gases hampered mining and rendered men unconscious or ill.

One commentator judged mine fires "the most appalling calamities imaginable. An explosion is over in a moment, but the slow suffocation by gas and smoke is a lingering death of the most agonizing kind." Mine fires frightened the men underground; they remained unaccountable and uncontrollable. They killed indiscriminately; the skilled and conscientious perished with the careless.[35]

> Clink! Clink! Clink!
> The song of the hammer and drill!
> As he toils in the shaft, in the stope or raise,
> 'Mid dangers which lurk, but elude the gaze,
> His nerves with no terrors thrill.

Though romanticized, Alfred C. King's stanza on the unseen dangers in the stopes and raises gives an image of the miners' lives. Fire and water claimed their victims, but the vision of slow death by entombment proved even more frightening and pervasive.

> Clink! Clink! Clink!
> For the heart of the miner is strong and brave;
> Though the rocks may fall, and the shaft may cave
> And become his dungeon, if not his grave,
> He braves every thought of ill.[36]

Merely an element of the macabre for most people, accidental interment was for the laboring miner a real possibility. Though miners knew the signs of danger in the stopes and drifts, they lacked the authority or means to correct the problems. Inadequate or deteriorating timbers, prolonged rain, subsurface water, earthquakes, and severe explosions were the most common instigators of cave-ins. Visible warnings of impending disasters were common; rotten timbering and wet ground were the most obvious, but often attentive miners could also notice or hear the shifting of the ground before the collapse occurred.[37]

---

[35] Williams, "Miner and Perils," p. 432.

[36] King, "The Miner," *Mountain Idylls*, p. 46.

[37] Cave-ins often resulted from careless or improper mining practices. For discussion of cave-ins and their causes, see Lord, *Comstock Miners*, pp. 217–219, 401; Henry C. Morris, *The Mining West at the Turn of the Century*, ed. C. S. Lewis, pp. 11–12; "Mining Letter," *Rawlins* (Wyoming) *Record*,

Psychological torment, injury, and death accompanied cave-ins. The dead never saw the cold rationalizations of their employers' agents, nor did they see the mourners lined up behind the draped hearses. They could find no comfort in J. C. Dickey's explanation of one cave-in: "It was one of those things that are bound to happen where so many men are working. This was a bad piece of ground, and has been for the past four months, and there was special care taken to select good timber, and it was found after the accident happened that the timbers could not have been put in better shape than they were."[38]

Frank Crampton was trapped in a Utah mine by a cave-in. Crampton later explained that the disaster followed an earlier powder-house explosion, which combined with water seepage and poor timbering to precipitate the cave-in. As the twenty victims surveyed the damage, they collected their supplies of food, candles, and snuffs, deposited other belongings in their lunch buckets, and prepared crude accommodations for their anticipated stay underground. They converted powder cases into makeshift toilets and established a routine that included tapping signals on the compressed air line.[39]

The trapped miners at first maintained their optimism in the face of this adversity by singing and joking. On the second day of their interment, they faced their first crisis, when their food spoiled, poisoning them and thereby heightening their discomfort. When the repercussion from a blast set off by their rescuers extinguished their candle, Crampton later remembered, ". . . it was a nightmare and far from unreal." Since their matches had become soggy, they could not relight their candle, so they suffered on in the dank darkness, unable to recapture their former optimism. For ten days the miners awaited rescue, and, when it finally came, they were all suffering from malnutrition, exhaustion, and prolonged exposure. Crampton spent almost two months recuperating, while the immigrants, less fortunate, were probably forced to return to work as soon as they could move. The episode illustrated the comradeship and *esprit de corps* among

October 18, 1905; Mabel Barbee Lee, *Cripple Creek Days*, pp. 254–256; Crampton, *Deep Enough*, pp. 106–120.

[38] J. C. Dickey to Irving T. Bush, Central City, January 15 and January 20, 1898, Gregory Consolidated Mines Papers, CU.

[39] Crampton, *Deep Enough*, pp. 107–110.

miners, as Crampton's three closest friends, Jack Commerfield, John Harrington, and Michael Sullivan, all joined in the rescue attempt, even though none had been employed in the mine where Crampton worked and only Commerfield had been in Utah.[40]

Yet all the physical dangers discussed so far pale before the hazard of quartz particles, the implacable foe of the working miner. Miners' consumption, or silicosis, struck without warning; it was a cumulative disease resulting from prolonged contact with microscopic quartz dust. Its symptoms, like those of emphysema, appeared after the disease had reached an advanced and often irreversible stage. Gradually, scar tissue formed in the lungs and impaired their functions. Nicknamed "miners' con," the ailment varied in severity depending on the length of exposure to the dust, the regularity of this contact, the intensity and nature of the work performed in the dusty environment, and the general working conditions. For example, men working in poorly ventilated raises or near dry-drilling proved most susceptible to the disease; thus drilling at the face, a job of high status, was also one of the most dangerous.[41]

While miners' con could kill by itself, it more frequently contributed to other fatal pulmonary illnesses, specifically pneumonia and tuberculosis. No accurate statistics on the prevalence of consumption are available for the years before the twentieth century. In their 1921 study of Butte, Montana, Daniel Harrington and A. J. Lanza found that 42 percent of the volunteers tested had the disease. Cleve E. Kindall, a U.S. Bureau of Mines official, found the situation worse in Tonopah. "From memory I will say at least 90% of the men examined are affected with the dust, an appalling rate. We certainly will be S.O.L. for any future work there after the report reaches them [the operators]."[42]

Miners' consumption ordinarily struck the older men, the family men. In Central City, Colorado, three separate Cornish families lost one or more adult miners to it. In Delamar, Nevada, the dust proved particularly damaging to lung tissue because of its high quartz con-

---

[40] Ibid., pp. 110–120, quotation from p. 112.

[41] Harrington and Lanza, "Miners' Consumption," pp. 7–9; E. Smith, "Mine Safety," p. 4.

[42] Harrington and Lanza, "Miners' Consumption," p. 11; Cleve E. Kindall to E. D. Gardner, January 26, 1922, E. D. Gardner Papers, WHRC.

tent. The situation there received widespread attention because the quartz dust was also released in the milling process, and, since the mill was upwind from the town, all residents were exposed to the floating particles. While the milling operations in Delamar were eventually cleaned up, miners there continued to work exposed to high doses of quartz dust.[43]

Consumptive miners ordinarily gave up their strenuous occupation. Their weakened lungs could not handle the pulmonary requirements of straining bodies. If the victims were lucky, they might find or be given work on the surface. However, few corporation officials showed the consideration and concern for a long-time employee that W. Rowland Cox displayed toward Sam Northey, one of consumption's victims. "I am very sorry to hear of Sam Northey's condition and believe he should move out of the mountains, also it will probably be necessary for him to have work on the surface hereafter, and we should try to place him somewhere else. I think we owe him this consideration." More often the diseased drifted away, seeking employment wherever they could. As Cox pointed out, the best treatment demanded an end to mining and departure from the rarified mountain atmosphere. Other consumptives, unable to leave, struggled on, living from seizure to seizure and knowing that the next might be the last. Mabel Barbee Lee vividly remembered the fits of coughing and gasping that afflicted her father, a miner and prospector in Cripple Creek. "A sudden fit of coughing seized him, the first he'd had for many days. He leaned far over, clutching his throat and choking for breath. It seemed as though he would never stop. His face turned purplish red and beads of perspiration stood out on his forehead." According to her, friends and families of consumptive miners tried to keep the affliction quiet, as there was a stigma attached to it. The disabled and dying certainly were present in the mining towns, but they were seldom mentioned and their disease was rarely discussed.[44]

As if these hazards were not enough, miners also worked in un-

[43] Todd, *Cornish Miner*, pp. 161–162, 168–173; John M. Townley, "The Delamar Boom: Development of a Small One-Company Mining District in the Great Basin," *Nevada Historical Society Quarterly* 15 (Spring, 1972), 10, 11–12.

[44] Cox to Wheelock, January 11, 1917, Blumenthal Papers, UNM; Lee, *Cripple Creek Days*, pp. 63, 68–70, 184, 220–222.

sanitary surroundings. Horses and mules were used in tramming; the United Verde Mine of Jerome, Arizona, used animals until well into the twentieth century. In Utah one mining company was still using horse-drawn tram cars as late as 1912. Excrement littered the drifts and crosscuts of the mines. The miners themselves were careless; scraps from hundreds or even thousands of lunches and suppers produced an unpleasant aroma, which was further worsened by human ordure. The odors were intensified in the poorly ventilated depths and confined quarters of the metal mines. More serious than the smell were the dangers from diseases thought or known to have been transmitted through excrement: hookworm, typhoid, tuberculosis, and dysentery. The gradual replacement of animals with electrically powered haulage systems eliminated part of the problem, and the U.S. Bureau of Mines urged companies to provide some type of underground latrines for their employees. In a report entitled "Underground Latrines for Mines," Joseph H. White of the Bureau of Mines argued that mines were no different than workshops and factories and concluded that defilement was both unnecessary and objectionable. He advocated the use of underground latrine cars.[45]

Though underground conditions were often unsanitary, their full effects on miners are not known. Did the organic waste provide a breeding ground for microorganisms? To what extent were communicable diseases spread from individual to individual in the stopes and drifts? Thorough analyses were not completed when the victims and data were available, and, as in so much that is associated with mining and its dangers, the miners themselves rarely took the time to explain their situations.

Miners, it is said, accepted their fate with impassive stoicism. Superstition, however, is probably a more accurate term for their attitude toward the uncertainties of their occupation. Where caution failed to ward off catastrophe, men turned to supernatural causes and spiritual intervention as explanations for the course of human events. In time, certain activities and occurrences took on symbolic meaning and became recognized portents. The horseshoe was a common sign of good luck. With its points upward, it not only appeared at the mine

[45] Joe [Manwaring] to E. D. Gardner, January 24, 1912, Gardner Papers, WHRC; Joseph H. White, "Underground Latrines for Mines," U.S. Bureau of Mines, *Technical Papers* 132 (1916), 5–14.

entrance, but also graced the entrances to tunnels, drifts, and levels. Some miners actually refused to work in mines that lacked horseshoes. Playful wags' gleeful observation that some of the worst mines had more than their share of rusting horseshoes seemed never to diminish the superstition. While horseshoes were supposed to bring good luck, whistling underground was considered the worst possible omen. Supposedly, whistling drove out the good luck spirits. Some miners further believed that it scared the ore away. While they regarded women in the mines as bad luck, American miners commonly believed that if they met a pretty woman on their way to work, they would enjoy good luck for a week; if the lady smiled, they expected a full month of good fortune. Other portents of impending disaster included clothes that fell from their hooks in the change houses and the announcement by a miner that he was working his last day. A candle that went out three times or fell from a tunnel wall, supposedly indicated that another man was calling on the wife of the one who witnessed the sputtering. There were common taboos; superstitious miners refused to change their clothes once they had put them on for the day and often would not return to the mine to retrieve a forgotten article of clothing or lunch pail. Both actions were believed to summon bad luck.[46]

Some of the miners' customs were quaint and humorous. For example, Cornishmen believed that by bringing their mules to the surface on Christmas Day, they would ensure their own good health in the coming year. Cornish and Mexican miners also believed that little people lived in the mines. The Cousin Jacks called the wee folk tommyknockers, unseen companions who guided them to the good ore and always warned them in time of danger. Though no Cousin Jack, Frank Crampton believed the story and even credited the tommyknockers with once saving his life.[47]

Mexican and Indian miners believed that mines were inhabited by a dreadful "step devil." Usually described as a dwarf with abnormally long arms, the creature struck in times of disaster. It

[46] "Miners' Superstitions," *Mining and Scientific Press* 38 (1879), 250; "Miners' Superstitions: Whistling Underground," *Mining and Scientific Press* 44 (1882), 54; Jack Foster, ed., "Colorado Question Box," *Denver Rocky Mountain News*, September 28, 1958, clipping in Mining File, WHRC.

[47] Crampton, *Deep Enough*, pp. 86, 87.

climbed the underground ladders ahead of the frightened miners and pulled itself upward with its hands while stamping off the lower ladder rungs with its feet. The Indians, who used notched logs rather than ladders, pictured it as a dwarf monster with long toe nails with which it gouged out the foot notches as it ascended. The spirit of the step devil was perpetuated in other myths of ghosts thought to destroy the miners' means of escape in times of crisis.[48]

Miners displayed an unwillingness to work in stopes or drifts where companions had recently lost their lives. They were particularly anxious to retrieve the remains of men killed underground lest their unclaimed spirits haunt the workings. Strange sounds in abandoned portions of mines were transformed by the superstitious into fearful apparitions that stalked unsuspecting miners. Practical jokers preyed on such fears. William Wright recorded one instance of an apparent ghost in the 700-foot level of the Comstock's Ophir Mine. In fact, several miners had masqueraded as the lonely spirit of a long-dead miner and successfully frightened one investigator. The mine's unsuperstitious electrician decided to discover the cause of the disturbances and finally cornered one of the jokers, who retorted angrily: "Well, *you* don't scare worth a cent!" This mysterious ghost then disappeared, never to be heard from again.[49]

Misfortune and tragedy occasionally led miners or their families to local spiritualists, who claimed to be able to divine the future and understand the workings of fate. Adeline Waldorf turned to Mrs. Sandy Bowers, a Washoe seeress, after she lost one of her children. Rich and poor alike consulted Mrs. Bowers, who claimed that she saw everything from impending mine fires to the location of untouched deposits of high-grade ore.

Some men believed that lightning was the best guide to large mineral deposits, even when the bolt killed a man in the process. Indulging in yet another superstition, men like Jonce Barbee and John F. Campion of Colorado believed that the very names of their mines and claims would bring good luck. Thus Jonce Barbee changed the name of his Cripple Creek prospect to "something patriotic," the

[48] "A Mexican Mining Legend," *Mining and Scientific Press* 44 (1882), 326.

[49] J. C. Dickey to Irving T. Bush, Central City, January 24, 1898, Gregory Consolidated Mines Papers, CU; Wright, *Big Bonanza*, pp. 251–253.

Columbia, in the hope that it would change his luck. Leadville's mining king John Campion believed that the names of animals in the deer family were lucky, so he named his mines the Ibex, Deer, Moose, Antelope, Kidoo, and Bison. Whether the names actually made any difference is academic, but Campion did enjoy good fortune once he switched to this unusual nomenclature, and he, at least, attributed it to the names.[50]

Miners' superstitions, like superstitions generally, attempted to account for the inexplicable and irrational events in human experience. Few occupations were as risky or as uncertain as laboring in the mines. When men perished or were disfigured by almost any happenstance—from striking a protruding rock with their skulls to walking beneath a discarded and rapidly falling pebble—they sought explanations that seem as illogical to modern minds as their perils must have seemed to them. Accidents had always been part of mining, although the introduction of new technology occasionally changed their character. The risks of mining inevitably influenced the aspirations of the miners, and helped shape the individual miner's attempt to earn a living for himself and his family.

[50] John Taylor Waldorf, *A Kid on the Comstock: Reminiscences of a Virginia City Childhood*, ed. Dolores Bryant Waldorf, pp. 18–19; Wright, *Big Bonanza*, p. 62; Parsons, *Journal*, October 24, 1880, p. 176; Lee, *Cripple Creek Days*, p. 165; Carlyle C. Davis, *Olden Times in Colorado*, p. 196.

6.

Earning a Living by Mining

THROUGHOUT this era, 1860–1920, mining was a largely industrial occupation, and the typical miner, a wage earner. Initially, both wages and living costs were artificially inflated by the scarcity of labor and the isolation of the mining communities. As the districts matured, labor became more abundant and towns less isolated; then mine-owners, especially those involved in corporate consolidations, began to press for wage reductions. Lower wages spurred miners' interest in alternatives to simple wage earning, especially since the cost of living in mining towns remained relatively high.

Food, services, and shelter were expensive; almost everything had to be transported into the districts. Initially, communities depended upon pack trains and horse-drawn freight wagons. In towns that had producing mines with sizable ore deposits, railroads soon comple-mented the other forms of land transport and reduced transportation costs. Even then, however, freight rates remained high.

Board for single men varied from a low of five dollars a week to a high between two and three dollars per day. The lower rates pre-vailed both in cities like Cripple Creek and in remote camps like St. Elmo, Colorado. Higher prices were common in all towns during the initial rush. Edward R. Howe reported that he was paying seventy-five cents per meal in the roaring camp of Alba City, Utah, in 1872. Henry C. Morris found comparable rates in the boom camps of early twentieth-century Nevada. In Tombstone, Parsons paid fifty cents per meal. Monthly board payment of twenty to ninety dollars was the typical range until the inflation accompanying World War I.[1]

[1] Leo J. Keena, "Cripple Creek in 1900," p. 8, Leo J. Keena Papers, State Historical Society of Colorado, Denver (hereafter cited as SHSC); Bessie Launder Richards, "Mining Town Memories—Colorado and Mexico," type-script of interview conducted by Mel Erskine, pp. 7–8, Bancroft Library, University of California, Berkeley (hereafter cited as BL); Edward R. Howe to

Food for families ran about a dollar a day, lower than for single men because wives prepared the meals and cleaned up afterward. The diets were as varied as the provisions and the family budgets allowed. Starches, fruits, meat, and eggs were normally available. The price of beef fluctuated between fifteen and seventy-five cents per pound. Coffee, a staple, usually cost between seventy-five cents and a dollar a pound. Flour brought between ten and fifteen dollars per hundredweight.[2]

Added to the cost of food was the expense of housing or lodging, and again the marked variations from place to place reflected the state of development of the individual communities. Under the best circumstances, men paid no more than ten dollars per month for a room. Houses and two-room tent-cabins rented for between fifteen and twenty dollars in producing camps, and rates fell precipitously after labor strikes or in declining regions. Land prices and building costs soared in boomtimes. Town lots often doubled in value within a few months, only to fall as rapidly should production cease. As a rule, though, lodging added an additional ten to twenty dollars to the miner's monthly cost of living.[3]

Dr. J. Hasting, September 25, 1872, Edward Robbins Howe Papers, BL; Henry C. Morris, *Desert Gold and Total Prospecting*, p. 31; idem, *The Mining West at the Turn of the Century*, ed. C. S. Lewis, p. 52. "The Dollar Is Shrinking," *Miners' Magazine* 18 (June, 1917), 6, reported that costs of food and other necessities in Leadville had increased as much as 70 percent, while wages had risen only 14–16 percent.

[2] Information on commodity prices and the cost of living was compiled from Bureau of Labor Statistics of the State of Colorado, *Fifth Biennial Report, 1893–1894*, p. 72; Richards, "Memories," p. 6; Thomas D. Sanders and Charles C. Stemmer, "Hardships and Perils on the Western Frontier or Fifty Years of Pioneering," p. 112; Journal of William Cowper Wood, July 23, 1868, Henry Ellsworth Wood Papers, Huntington Library, San Marino, California (hereafter cited as HL); John Shumer to J. Wilbur, Star District, Utah, November 26, 1870, Utah Mining Papers, HL; Barton Wood Currie, "Housekeeping in Alkali Land," *Good Housekeeping* 49 (August, 1909), 119–120; Diary of George O. James, November 24 and December 27, 1860, Western History Department, Denver Public Library (hereafter cited as DPL).

[3] Bureau of Labor Statistics of the State of Colorado, *Fifth Report*, p. 72; "Stay Away from Arizona," *Miners' Magazine* 18 (February, 1917), 6; Diary of E. D. Gardner, June 2, 1907, E. D. Gardner Papers, Western History Research Center, University of Wyoming, Laramie (hereafter cited as WHRC); James McAuliffe to Thos. Waters, Sutro, Nevada, May 12, 1879, Thomas Waters Papers, BL.

Other expenses included clothing, fuel, water, milk, insurance, and interest on borrowed money. In 1896, the Colorado Bureau of Labor Statistics estimated these expenses at twenty-five dollars per month for Leadville miners. Thus, without calculating incidentals like gambling, union and fraternal dues, drinking, whoring, and theater-going, miners spent between $55.00 and $125.00 per month on necessities. This standard of living required a $2.50 daily wage. In general, high prices for basic necessities were balanced by correspondingly high wages for miners. But as districts matured or declined, the costs of goods and services stabilized or declined, bringing pressure for wage reductions.[4]

Wage work was the simplest and most common form of employment. Most miners remained in wage-paying jobs as long as they remained healthy and their work satisfied their shift bosses. As an alternative to simple wage earning, though, some miners chose to work on either contracts or leases. In contract mining, an individual or a group of men agreed to do a specified amount of work at a standard rate. Miners tunneled, drifted, or sank shafts at fixed sums per foot or at a stated price for so many feet or fathoms. Usually contracts were let to the lowest bidders and contained clauses that clearly established the quality and quantity of work to be done, the payment procedures, and the materials that the miners were expected to furnish. Some miners favored contracting because sustained hard work meant higher daily earnings, while some companies preferred it because contractors worked harder and faster than their wage-earning counterparts.

Lease mining differed from both contract and wage work in that the leasers assumed greater risks and expected greater dividends. Leasing was most common in young camps or in declining districts where owners and companies sought a limited-risk return on their investment. The leasers assumed all the costs and obligations of the mining and paid the owners anywhere from 10 to 60 percent of the gross receipts from the ore removed. When accompanied by a guarantee bond, leasing was the closest thing to an assured return for the owners. On the other hand, miners holding leases could make money if they had a good agreement and encountered a solid pocket of high-grade ore. Leasing was the principal way in which working miners

[4] Bureau of Labor Statistics of the State of Colorado, *Fifth Report*, p. 72; Diary of Ella Baily, September 28, 1869, SHSC.

accumulated small fortunes, an uncertain but tantalizing avenue for economic mobility. Miners ordinarily leased undeveloped properties or levels of larger establishments, for the leasing of an entire property required more capital than working men could raise.

The most typical miner, though, was the wage earner. With an annual cost of living between $660.00 and $1,500.00, he received daily wages varying from $3.00 to $5.00, an annual 300-day earning of $900.00 to $1,500.00. With twelve-hour shifts and $3.00-a-day pay, wages came to $0.25 an hour. As eight- and ten-hour shifts became more common in the early 1900's, minimal hourly wages rose to $0.30 or $0.375. By comparison, manufacturing workers did not exceed the $0.25 per hour rate until 1902 and were still below the $0.375 figure in 1917. The average annual earnings of workers in all industries did not exceed $900.00 until 1918, and farm labor did not surpass this figure at any time before 1920. As one miner remarked, "Any man who was broad in the shoulder and narrow between the eyes could get a job as a mucker—and his $3.00 a day for eight hours work must be looked at against a background of the 10 cents an hour wage for a ten or twelve hour day which prevailed in most of the country for similar work in industry."[5]

There were some exceptions to the prevailing wage scales. Notably, the older camps of Colorado's eastern slope paid lower wages. The $2.00 per day and board paid by the Stevens Mining Company of Georgetown was rock bottom, but other operations in Georgetown, Brownsville, Idaho Springs, Central City, and Black Hawk also paid less than $3.00. The region's proximity to Denver and the special costs of processing its notoriously refractory ores explain these low rates. Nonetheless, both Lynn Perrigo and Arthur Todd have concluded that Central City's Cornish miners lived fairly well on daily wages of $2.50. Some companies expected miners to provide their own equipment, and this altered real wages.

[5] R. Orchard Old to Subcommission on Mining of the U.S. Industrial Commission, Georgetown, April 22, 1899, Robert Orchard Old Papers, SHSC; Wyoming State Mine Reports, Grace Raymond Hebbard Papers, WHRC; Bureau of Census and Social Science Research Council, *The Statistical History of the United States from Colonial Times to the Present*, pp. 91–92; Keena, "Cripple Creek," p. 2, Keena Papers, SHSC. The U.S. Bureau of Mines used a 300-day year as its basis of comparison for accidents and wages, because that

Wage variations angered most miners, who believed that mining work was fundamentally the same everywhere and that wages should reflect this similarity. However, sometimes miners themselves initiated lower wages. For example, in Turquesa, New Mexico, S. A. Larock offered his services to L. Bradford Prince for $2.75. While noting that miners at the nearby Cash Entry earned from $3.00 to $3.25 daily, Larock proposed to accept lower wages because he was then unemployed and considered the work at Prince's Baltimore Mine "not very hard." For the most part, though, the wage scale was relatively constant over this sixty-year period throughout the region; nascent camps paid $4.00 to $5.00 and mature communities, about $1.00 less.[6]

Unlike wage rates, which were notable for their constancy, the pay for contract work varied considerably, even within the same community. Rates depended on the type of work undertaken, the geological character of the ground worked, the labor supply, the respective obligations and reputations of contractors and companies, and the amount of work and the duration of time for which the contract was drawn. For management, contracting combined the advantages of the piecework and speed-up systems, and though one twentieth-century miner referred to it as the "contract gyp system," he conceded that it was practically the only way in which miners could earn a few extra dollars for their families or build up a stake for the future.[7]

---

allowed for injuries, holidays, illness, and layoffs—all integral parts of the miner's working lot.

[6] Richard A. Pomeroy to Nat Harris, Georgetown, February 1, 1878, Stevens Mining Company Papers, DPL; Elmer Faunce to S. G. Calkins, Georgetown, January 2, 1874, William A. Hamill Papers, DPL; Payroll of Gregory Consolidated Mines for December, 1897, Gregory Consolidated Mines Papers, Western History Collection, University of Colorado Libraries, Boulder (hereafter cited as CU); advertisement for miners, *Central City Daily Register*, May 14, 1871; Lynn I. Perrigo, "The Cornish Miners of Early Gilpin County," *Colorado Magazine* 14 (May, 1937), 94; Arthur Cecil Todd, *The Cornish Miner in America*, p. 165; S. A. Larock to L. B. Prince, January 10, 1890, L. Bradford Prince Papers, State Records Center of New Mexico, Santa Fe (hereafter cited as RCNM); Employee Time Slips of the B.O.B. Mining and Milling Company, New Mexico, January–December, 1900, Holm O. Bursum Papers, Zimmerman Library of University of New Mexico, Albuquerque (hereafter cited as UNM).

[7] Seth Cook to George O. Whitney, Virginia City, January 10, 1862, Sierra Nevada Silver Mining Company Papers, HL; Charles A. Smith to Gerald

Contracts were a popular way to handle necessary dead work; for example, Cecil C. Morgan, superintendent of Georgetown's Colorado United Mining Company, relied on them for that. Agreements on gold and silver mines often contained clauses that voided the deal should high-grade ore be found. For example, the notice about the terms for contract work issued by Fred G. Bulkley, manager of Leadville's Ibex Mine, specifically stated that "contracts will be let for fifty feet of work at a time, unless pay ore be sooner encountered, in which case the contract will terminate when pay ore is struck." Management seldom trusted unsupervised miners to extract precious high-grade ore. Because high-grading—slipping some rich ore into one's pocket—was tempting and even sanctioned by prevailing custom, owners insisted that contracts be scrapped when rich ore was struck.[8]

Managers offered contracts to their miners for one of two reasons: speed or economy, and ordinarily the latter. For example, in 1902 Uri B. Curtis proudly reported from Butler, Nevada: "My contract for sinking the shaft I was able to make at $5.00 a foot, the party doing the work to furnish the powder, fuse, candles, and caps; I, to remove the dirt. Then I made a sub-contract with him to take out the dirt at two and a half a shift. It was very much cheaper than I would have been able to have done it."[9] In 1871 John Shumer of Utah's Star District put the matter succinctly when he noted: "I will make arrangements as soon as possible, to go to work on the 'Coyote' either by letting a contract for sinking a shaft, or by hiring men to help me sink it. By hiring men, it will cost about $10 a foot, and if I can let a contract for less than that I will do so."[10] Six years earlier Thomas M. Yenkes had seen additional economy in contract work because it provided incentive for his miners in Pima County, Arizona. He found "it cheaper whenever it can be done to task the men or give

Sherman, Miami, Arizona, March 28, 1922, and Sherman to P. G. Beckett, May 9, 1922, Frank Ayer Papers, WHRC; interview with Albert J. Frantz, Denver, April 6, 1972.

[8] Diary of Cecil C. Morgan, February 8, 1878, DPL; Notice of Contract Work, Ibex Mining Company, Fred G. Bulkley, manager, April 20, 1899, John F. Campion Papers, CU.

[9] Uri B. Curtis to Clyde A. Heller, Tonopah, February 16 and June 29, 1902, Daniel Moreau Barringer Papers, WHRC.

[10] John Shumer to J. Wilbur, Star District, Utah, July 26, 1871, Utah Mining Papers, HL.

them a small contract to finish out for so much. They work at it with a better relish." The results of contracting clearly pleased both engineers and managers. Writing in 1922, Gerald Sherman assured P. G. Beckett, general manager of Phelps Dodge, that the "advance was immediately increased about 50%, with the contract system."[11] It is little wonder that contracting remained a popular mode of employment.

For the miner, it was critical to analyze the areas open to contract work and evaluate the terms carefully. On rare occasions, especially with undeveloped properties, miners and management jointly drafted the terms of the contract. For example, between 1885 and 1889, L. Bradford Prince received numerous contract proposals from miners in New Mexico's Cerrillos Mining District, sometimes in reply to local notices circulated by Prince or his representative, sometimes with no specific solicitation. Prince's properties, the Pinon, the Baltimore, the Open Sesame, and the Sleeping Beauty, were little more than prospects, and contracting allowed Prince to develop them without close attention to the work in progress.

Miners like S. A. Larock responded to contract notices because they needed work. Writing in January of 1885, Larock stated:

I would like very much to get the work as times are dull at present and not much doing. I will sink 11 feet for $100.00 or any number of feet between that and 50 for $10.00 per foot, pay such in 60 days after completion of work.

I would like to get your work on the Open Sessame [sic] also. I will down the tunnel in 12 feet farther for $100.00 and wait until first of January 1886 for pay.

However, as his letter of August 31, 1887, indicated, even Larock was unwilling to take a contract that would not pay for his effort. "Yours of the 29th at hand, my bid of $6 per foot included the ore I would take out. It cannot possibly be done for that price as it is worth at least $8 per foot. However if you prefer to keep the ore I will down it for $7.50 same as last year.[12] In response to Prince's notice for contractors in 1889, Joe Rossa was even more specific. When Prince re-

[11] Thomas M. Yenkes to H. D. Bacon, April 15, 1865, Henry Douglas Bacon Papers, HL; Sherman to Beckett, May 9, 1922, Ayer Papers, WHRC.

[12] S. A. Larock to L. B. Prince, Turquesa [N.M.], January 10, 1885, and August 31, 1887, Prince Papers, RCNM.

quested proposals for the amount of work to be done for a set $100, Rossa replied that he would undertake "15 feet on drift 4' x 6' the same to be competed [*sic*] within 60 days from the time of agreement with you for the contract. You to acknowledge the work when completed in a workmanlike manner and pay on or before the 10th of Dec 89."[13]

The flexibility of the agreements made with Prince contrasted sharply with the rigidity in more established mines, where management unilaterally assigned the respective responsibilities of company and contractors. Forced to accept the company's terms or lose the contract, the miner and his partners had to determine whether or not they could make enough money to justify their effort. Leadville's Ibex Mining Company minutely specified the respective obligations of the company and the contractors.

THE IBEX MINING COMPANY WILL FURNISH all tools, ammunition, lights, mine timbers, track rail, spikes and ties; and air for ventilation, where now furnished.

THE CONTRACTOR WILL FURNISH all necessary labor, including rock breaking, cleaning up and tramming, and tracklaying, and will place all necessary timbers.

There must be four signers to each bid, and all must be working partners. No substitutes will be allowed without the consent of the Superintendent.

Contractors will be under the direction of the Company at all times, as to timbering and as to driving straight ahead or turning, and they will be subject to all the rules of the mine the same as other miners.

Contractors will be required to save separately all low grade or shipping ore broken.

Work must proceed continuously to the completion of the contract, with at least two regular shifts.[14]

Contracts typically specified the rate at which the sinking, drifting, or stoping was to be done, the feet or fathoms of advance expected, the obligations of the miners and the company, and the payment procedures. For example, in the 1870's the Manzanita Mining Company of Colorado's San Juan region specified the conditions under which it would let tunnel contracts.

[13] Joe Rossa to Prince, Turquesa, August 9, 1889, Prince Papers, RCNM.
[14] Notice of Contract Work, Ibex Mining Company, April 20, 1899, Campion Papers, CU.

The tunnel is to be straight seven and one-half feet high by six feet wide in the clear or inside the timbers, and have a uniform rise or grade of seven inches in each fourteen feet, to be well and securely timbered where timber is needed, uniform with that portion already finished. Contractors will be required to work at least twenty shifts each week of two or more men to each shift, and to furnish themselves with everything except timbers and lagging car and lumber for cartrack, lumber and nails for air boxes and the necessary use of the Company's blacksmith-shop at the mouth of the tunnel. The work to be done in an expeditious and workman-like-manner under the daily supervision of the Company and to their satisfaction. Payment will be two-thirds of the amount due at each one hundred feet, the balance 60 days after the contract is completed.[15]

In addition, some contracts included provisions for the extension of a contract beyond the agreed-upon distance and time period. When companies wanted work done rapidly, the contract might offer an incentive payment. Leadville's Lee Basin Mining Company offered William Landers an additional $200 if he completed his fifty-foot sinking contract within thirty days. Penalty clauses were also part of most contracts. Ordinarily, miners were paid only after specified amounts of work had been completed to the company's satisfaction. Noncompliance with the terms of the contract meant that the miner forfeited his right to the contract and to payment for work already done. On the other hand, if a company failed to uphold its contractual obligations, the miner had little recourse. If he needed the money and believed that the company would eventually honor its debts, he could proceed with the work and hope for the best. His only other choices were to sue for noncompliance or to give up the contract.[16]

Contracting was a fine art; both the company and the individual miners sought to secure the best possible terms. A. D. Giles, a contract miner, made his self-interest clear to L. B. Prince in a letter of April 3, 1886. Responding to Prince's objections about Giles's inclusion of provisos to deal with additional contingencies, Giles noted, "The contractor is only bound to the performance of those things

[15] Manzanita Mining Company, "To All Contractors," [187?], Mining File, WHRC.

[16] Notice of Contract Work, Ibex Mining Company, April 20, 1899, Campion Papers, CU; contract between Wm. Landers and Lee Basin Mining Company, Leadville, November 6, 1886, and agreement between John H. Duhme and E. A. Quilbault, partners, and Lee Basin Mining Company, Leadville, December 6, 1888, both in J. J. Blow Papers, CU.

mentioned in his offer or proposition for work—and if anything *in addition* is to be done, he is entitled to additional compensation." Yet even having set forth his objections, he agreed to settle with Prince in a compromise that fell short of Giles's original demands.[17]

Occasionally both miners and their employers were entirely satisfied with the results of a contract. In the early twentieth century, Frank Crampton and his companions, John T. and Sully, took a drifting contract in one of Cripple Creek's mines. Maintaining a rigorous schedule of twelve-hour shifts, the friends completed their 300-foot contract in six weeks and left Cripple Creek with a sizable stake. However, their routine allowed little time for leisure or rest; Crampton noted that others who tried to follow their schedule soon abandoned their contracts in desperation. Even Crampton admitted that "often I wanted to call it deep enough, but I didn't let the boys know about that."[18]

Not all contracts were so satisfactory; the relations between managers and the contract miners, as well as those between contractors and their employees, were often strained by the terms of the contracts or by the poor performance of the work under contract. Employed as a day laborer by a contractor named Gallup, George Parsons found the work toilsome, his coworker incorrigible, and his pay delinquent. After almost a month of intermittent labor, Parsons finally settled his last claim against Gallup, who gave him his own mining tools to cancel a thirteen-dollar debt. Other miners too were aggrieved by not being fully compensated for work done under contracts. One Virginia City miner, when denied payment on a previous contract, countered by refusing to continue work on a second one until fully reimbursed for his earlier work. He threatened to work as slowly as possible until the company met his demands, whereupon the company merely offered the contract to another group of miners. Although the company had to accept a slightly higher bid, the malcontent had been dealt with and his back wages remained unpaid. In Georgetown, Colorado, a group of Cornish miners stopped work on a contract be-

[17] A. D. Giles to L. B. Prince, Turquesa [N.M.], April 3, 1886, Prince Papers, RCNM.

[18] Frank A. Crampton, *Deep Enough: A Working Stiff in the Western Mine Camps*, pp. 45–46.

cause progress was so slow they were not earning the equivalent of a daily wage. Even though the men had drifted six feet through the hardest rock in the region, the resident manager affirmed that "if they quit I shall not pay them anything for what they have done."[19]

Even more reprehensible were the activities of corporations like Utah's Apex Mining Company. According to Edward R. Hastings, Apex officials granted contracts to high bidders in return for kick-backs. The stockholders would have been justly angered at such shenanigans, but supervision of corporate officials was often lax, and individual miners rarely took the pains to bring such abuses to light.

Still other difficulties beset contractors. The pressure of the work and life's distractions added to the difficulties of most contractors. Albert B. Sanford reported that two Gunnison County miners abandoned their contract when one refused to allow his partner to fish while he himself continued to work at the mine.[20]

Contract mining differed from wage mining in that contracting was financially precarious and required both greater expertise and more personal responsibility. Miners had to estimate their expenses accurately, and they had to judge what might lie behind the next shot. Contracting posed the same problems for individual laborers that mining in general posed for investors and corporations. Contractors were not assured of earning anything, but if they were fortunate, persevered, and possessed even a rudimentary knowledge of mining and geology, they might earn a grubstake.

In general, corporate supervision of contract mining was sporadic, leaving the miners to determine how much timber was needed or whether reinforcement was necessary. The term "workmanlike manner" must have weighed heavily upon miners' minds; it was a two-word loophole that favored management. Furthermore, any accident might incapacitate the contractor. He then not only stood to lose the

[19] George Whitwell Parsons, *The Private Journal of George Whitwell Parsons*, March 16 and 26, and April 15, 1880, pp. 104, 107, 113; Seth Cook to George O. Whitney, Virginia City, January 10, 1862, Sierra Nevada Mining Company Papers, HL; Elmer Faunce to S. G. Calkins, Georgetown, March 8 and March 14, 1874, Hamill Papers, DPL.

[20] Edward R. Hastings to E. D. Gardner, Bingham Canyon, October 4, 1911, Gardner Papers, WHRC; Diary of Albert B. Sanford, July 28, 1881, Albert Byron Sanford Papers, SHSC.

pay for work he could not do but might well forfeit earnings he had already accumulated. Yet withal, contracting was not as risky as leasing, the supreme gamble.

In a memoir for her children, Pearl O'Brien Colwell described leasing as "the greatest and most dangerous 'bug' of all—and that's where Colwell money went." However, the six laboring miners who took up a lease on Central City's War Dance Mine in 1908 might have contested this indictment of leasing. Unable to find wage work, C. O. Richards, Neil McKay, Rue and Charles Manhire, and John and Michael Klais leased the War Dance, with a $25,000 option to buy the property. They struck high-grade shortly after they assumed the lease, had taken an estimated $100,000 worth of ore from the property by November, and then sold the property to eastern capitalists for another $100,000.[21]

Leasing provided managers a limited-risk way to work either undeveloped or aging properties. In undeveloped properties, limited capital put up by the owner was coupled with the nonwage labor of miners; both the owner and his leasers gambled that paying ore would be found. If they struck ore, the miners might do very well indeed, but, no matter what, the owner's risk was minimized. One such arrangement permitted John Marshall Paul to explore some properties near Fairplay, Colorado. Under a lease agreement that guaranteed him 58 percent of the receipts on all milling ore, he kept four miners working from July to December of 1875, even though they encountered no paying ore. Paul received five months of work and did not pay a cent. Each miner had sacrificed about $450, assuming that each could have been paid at the rate of $3 per day over the course of the five months. It is unlikely that such unprofitable work would have been undertaken had Paul been paying wages.[22]

The communities, the miners themselves, and the owners all had reasons to want the old mines kept open. Once closed, mines deteriorated rapidly, which made reopening them costly and dangerous. The loss of jobs and income to the community, as well as the loss of profit

[21] Pearl O'Brien Colwell, "Pearl," p. 12, Alfred B. Colwell Papers, WHRC; clipping from *Denver Times*, November, 1908, Central City by-line, dated November 7, 1908, Thomas F. Dawson Scrapbooks, XXII, 185, SHSC.

[22] John Marshall Paul to Harry, December 10, 1875, John Marshall Paul Papers, SHSC.

George Whitwell Parsons was an early Tombstone miner and member of that town's vigilance committee. Photo taken in the early 1880's. Courtesy, Arizona Historical Society.

Cripple Creek in 1891 typified early-day scenes in hundreds of mining camps. Courtesy, *Mining and Scientific Press*.

Geographical realities dictated the development of Bingham Canyon, Utah. Courtesy, Utah State Historical Society.

The Comstock in 1876 shows the complex urban environment that "mature" mining towns provided miners and their families. Courtesy, *Mining and Scientific Press.*

These cabins near Chicago Creek, Colorado, where George A. Jackson made his strike in 1859, are typical of the clapboard construction found in Western mining camps. Courtesy, Western History Collection, University of Colorado Libraries.

A burro train, loaded with cable for a nearby mine, fills the main street of Telluride, Colorado, in a scene typical of early mining towns. Signs identify a furniture store, a restaurant, a drug store, "ladies and gents furnishing goods" (a dry goods store), and furnished rooms for rent. Courtesy, State Historical Society of Colorado.

This Camp Bird, Colorado, one-room schoolhouse was photographed in 1903. Courtesy, Western History Research Center, University of Wyoming.

This photo shows the dining room of the boarding house at the Bingham Canyon, Utah, Daly Mine. Courtesy, Utah State Historical Society.

Pictured here is an early twentieth-century boarding house at Bingham Canyon, Utah. Courtesy, Utah State Historical Society.

The damaged mill of the Camp Bird Mine mutely testifies to the 1906 snowslide and fire. Courtesy, *Mining and Scientific Press*.

Relief from the dangers of mining and mining camps came in sporting contests and athletic teams. This is the 1889 Silver Plume, Colorado, baseball team. Courtesy, Denver Public Library Western Collection.

Keno, combining elements of bingo and cards, was a popular game among miners. Courtesy, Utah State Historical Society.

Rock-drilling contests were the miners' distinctive sport. This action shot of a drilling team was taken in Goldfield, Nevada, at the Independence Day celebration of 1904 or 1905. Courtesy, Western History Research Center, University of Wyoming.

Mines spawned numerous local assay offices, such as this one at Utah's Ontario Mine. Courtesy, Utah State Historical Society.

John F. Campion's Ibex or Little Johnny Mine in Leadville was a turn-of-the-century "extractory." Courtesy, Western History Collection, University of Colorado Libraries.

Freight wagons like this one at a Utah copper mine hauled both supplies and mine ore for the early mining communities. Courtesy, Utah State Historical Society.

In the early twentieth century, steam shovels like those pictured at this open-pit mine in Bingham Canyon, Utah, began to replace underground miners, just as the railroad cars and iron horse pictured had already displaced ore wagons and their teams. Courtesy, Utah State Historical Society.

These early Colorado miners "double-jack" into the ceiling of a stope. The man wielding the hammer (sledge) must use an underhanded motion to strike the drill held by his companion. Both men are spattered with candle drippings. Courtesy, Western History Collection, University of Colorado Libraries.

Underground workers sometimes mined alone (single-jacking), sometimes in teams of two or three, as in this photo. Courtesy, Western History Collection, University of Colorado Libraries.

Hand drilling gave way to mechanization. These machine miners are adjusting their drill for operation in a stope. Courtesy, Western History Collection, University of Colorado Libraries.

Two machine miners prepare to begin work with compressor, drill, and electric light, as well as the traditional candle (stuck in the wall near the light). Courtesy, Western History Collection, University of Colorado Libraries.

This early Goldfield, Nevada, miner is descending into the Combination Shaft by means of a bucket. Courtesy, *Mining and Scientific Press*.

Horse-powered winches, such as this one near Leadville, Colorado, were an improvement over earlier man-powered bucket systems. Courtesy, *Mining and Scientific Press*.

Still more sophisticated were the extractories' cage systems. These miners and an ore car are about to descend at the shaft of Utah's Ontario Mine. The track running up to the cages, the miners' attire, and the lantern are all typical for the period. Courtesy, Utah State Historical Society.

The extractories built complex headframe and hoisting systems like this one at the Congress Mine in Congress, Arizona. Courtesy, *Mining and Scientific Press.*

The interior of Utah's Buckeye Mine hoist house shows the cable take-up spool, the control levers behind it, and the operating gauges above. Courtesy, *Salt Lake Mining Review.*

"*Hence take notice that on and after Sept. 16, 1901, anyone working in and around the mines, mills or power plants of the Cripple Creek district who cannot show a card of membership in good standing of some local union of the Western Federation of Miners will be considered a 'scab' and an enemy to us, himself and the community at large, and will be treated as such.*

"*By order of the Cripple Creek Executive Board of the Western Federation of Miners. "JOHN CURRY, President."*

A poster found tacked to a telegraph pole the day after the awful Independence disaster, when thirteen poor fellows found out what it meant to be "treated as such."

This union poster, a warning to nonunion miners, allegedly appeared after the Independence Station explosion in the Cripple Creek District, June 6, 1904. Courtesy, Western History Collection, University of Colorado Libraries.

Pictured here are Smuggler Union miners of Pandora, Colorado, in the early 1900's. Courtesy, Western History Research Center, University of Wyoming.

Only a man in overalls — this miner, whose clothes are stuffed with dynamite and fuse, is ready to descend into a shaft on the bail of a bucket. Courtesy, Western History Research Center, University of Wyoming.

to the owners, thus made it desirable to all parties to keep old mines open whenever possible. Writing in 1912, E. D. Gardner explained why owners preferred leasing as a method of keeping these older properties operating. First, leasers could fairly easily be found. Local miners, remembering that the property had once produced paying ore, believed (often incorrectly) that more ore must be present; they were therefore eager to lease. While the assumption that ore remained was not always wrong, Gardner noted, "It cost me money to find out the fallacy of that hidden ore business."

A second advantage to the owner of leasing older properties was that the leaser, not he, assumed most of the costs. Therefore, the owner was able with relatively little expense to avoid shutting down the mine and, perhaps, incurring high costs to reopen it later if technology or the market again made it profitable. Finally, as a further result of the low investment required of the owner, leasing property with a tonnage royalty arrangement meant that any ore mined would be pure profit to him, regardless of the fate of others. The leaser, not the owner, paid the overhead, so the owner's share of the find, however small, was just that much more than he would have made with the mine closed. Here was the closest thing to a no-risk agreement available to mine owners.

Though Gardner did not discuss the relative merits of leasing a mine that was still producing, A. J. Anderson argued that leasing was the best way to operate one such, New Mexico's Slayback Mine. The Slayback possessed small pockets of ore, but a considerable amount of work was required to unearth the pay dirt. Anderson, who in some situations had himself leased from other owners, concluded that "the leasing system, particularly where we reserve the right to mill the ore, is the best way to work this property, both because of its distance from here [the mill] and because of the way the ore occurs."[23]

Contracts spelled out the responsibilities of the "leasers" (lessees). For example, in its lease with A. J. Anderson and Company (then acting as leaser), the B.O.B. Mining and Milling Company set forth Anderson's obligations.

[23] E. D. Gardner to Joe, February 18, 1912, Gardner Papers, WHRC; A. J. Anderson to W. Rowland Cox, Mogollon [N.M.], July 1, 1913, Blumenthal Papers, UNM.

To lease the said premises for the full term or period and comply with the covenants hereinafter set forth: to enter upon and into the said mining premises and work the same in a manner necessary to good and economic mining, with due regard to safety, development and preservation of the said mining property as workable mines and to preserve and care for the improvements on the said mines, including the mill thereon: to at once commence work upon the said mining premises and to continue the same thereafter during the terms of this lease and to do at least one hundred and twenty (120) shifts of work thereon each and every month: to mine, ship or treat all ores which have sufficient value encountered in said work on said mines.

Subsequent paragraphs required Anderson and his associates, W. R. Harris and H. O. Bursum, to keep the mine in satisfactory condition and to do all the necessary work of mining: "timbering, sinking, drifting, stoping, etc." In addition, the B.O.B. Mining and Milling Company bound Anderson to all the financial obligations incurred in mining; no debts or liabilities were assumed or accepted by the B.O.B. Company. Anderson and his companions paid a royalty on the ore mined and permitted the B.O.B. Company or its designated representatives free access to the property. If Anderson failed to meet the specified conditions, the lease would be voided.[24]

Leasing offered more than the lure of economic mobility and instant riches. Leasers exercised some discretion in their hours of labor, and they selected their own working companions. Semi-autonomous, leasers enjoyed some of the intangible benefits of self-employment. In the early twentieth century, the Western Federation of Miners (WFM) endorsed members' leasing as a temporary means of guaranteeing work to union members. Furthermore, the WFM felt that when the union itself or its members leased property, wage and hour standards and the closed shop would be introduced. The leaders believed that a few successful experiments might induce capitalists to adopt similar practices. R. B. Moudy noted that both the president and the secretary of the Cripple Creek union operated successful leases prior to the labor troubles of 1903. In Leadville, too, union leaders encouraged leasing. At the 1902 WFM convention, individual

[24] B.O.B. Mining and Milling Company lease with A. J. Anderson & Co., March 2, 1912, Blumenthal Papers, UNM.

miners endorsed union leasing and even tried to prod the national organization into taking out leases.[25]

One of the crucial components of any lease was the royalty that the leasers paid to the owners, based on the volume of pay ore extracted from the ground under contract. Ordinarily the fee was set as a percentage of the receipts from the milled ore, but that percentage varied according to the obligations assumed by the owners. The lowest recorded owners' percentage was 10 percent of all profits derived from the property, an arrangement secured by Colorado's mining magnate David H. Moffat from the Wolftone Consolidated Mining and Milling Company in 1891. However, as a prominent promoter and speculator with his own assets, Moffat could obtain better terms than most laboring miners. Not all leases were for a percentage of the find; the Wolftone corporation leased its Agassiz claim to R. D. Lawrence, R. H. Hazlett, and Maurice Starne for the right to use the excess water from the Mahala Mining Claim. Here a corporation granted a lease not for a royalty but for a mining necessity, water.[26]

Sometimes even laboring miners could secure leases that guaranteed the owners only a 10 percent royalty on the gross receipts. Miners John Olvany, John Carpenter, and John R. Bennetts proposed such an arrangement to L. Bradford Prince in the 1880's. In 1904 Thomas B. Catron concluded a similar agreement with C. A. Cooper and G. H. Bibb, Silverton, Colorado, miners. They promised Catron 10 percent of the gross receipts from his Animas River lode properties, regardless of freighting or smelting charges. An 1891 Silverton agreement allowed deduction of charges for mining, freighting, and smelting the ore but required a 12.5 percent royalty on the remaining income. The Alma Gold Mining and Milling Company required fully 20 percent of the receipts after deducting charges for sampling, milling, and smelting its ore. In one agreement for its Clear Creek, Colorado,

[25] Ross B. Moudy, "The Story of a Cripple Creek Miner," *Independent* 57 (August 18, 1904), 381–382; report of W. E. Giese, November 24–November 26 [no year], Campion Papers, CU; minutes of the executive board, Western Federation of Miners, November 28, 1902, International Union of Mine, Mill and Smelter Workers Papers, CU.

[26] Minutes of board of directors meeting, August 14, 1891, Record Book of Wolftone Consolidated Mining and Milling Co., pp. 32–35, Wolftone Consolidated Mining and Milling Company Papers, SHSC.

property, Alma Gold exacted a royalty of 35 percent after milling, railroad hauling, and smelting, *"Provided, However,* that no royalty shall be paid or demanded hereunder until said second party has for the period of one calendar month earned in the working of said premises under the terms thereof for himself and all other men employed by him hereunder the sum of two dollars fifty cents ($2.50) per working day including in said earnings the sum of two dollars ($2.00) per foot for all drifting and upraising done by him upon said premises."[27] This agreement assured William Arthur one month's wages before it invoked the stiff royalty arrangement.

While most contracts guaranteed the owners a flat-rate return, some corporations and individuals were concerned lest the leasers strike a rich vein and earn inordinate profits the owners could otherwise have enjoyed. The best safeguard was the so-called sliding-scale royalty. As the earnings of the leasers increased, so did the percentage owed the owners. In 1892, Ernest L. Foster proposed to lease the Gilman Mine to miner Edward Shovin, in return for a 10 to 25 percent royalty on the gross receipts of the ore milled. Although Foster promised to be flexible, so that Shovin would receive fair compensation for his labor, the experiences of Henry E. Wood suggested that such agreements often worked against the leaser. Wood had signed a lease of Leadville's Lilian Mine, which included a sliding scale requiring a minimum royalty of 10 percent and, on all production in excess of $10,000 per month, a maximum of 60 percent. Striking a rich vein, Wood had to pay the maximum rate throughout the term of the lease.[28]

The early Cornish miners of Gilpin County, Colorado, preferred leasing or, as they called it, the "tribute pitch" or "tut-work" method.

---

[27] John Olvany to L. B. Prince, Cerrillos, N.M., February 17, [188?], Prince Papers, RCNM; Catron lease with Cooper and Bibb, December 15, 1904, Bibb, John M. May, John Carr, C. M. Cramer, Howard Vaughn, Charles L. Thayer lease with Alex Perino, Anton Esher, Tom Ferino, and Michael Ferino, June 8, 1891, both leases in Rennehan-Gilbert Papers, RCNM; Alma Gold Mining & Milling Co. leases with John Peterson et al., 1902, Robert J. Calhoun, 1903, and William Arthur, 1902, Edwin L. Regennitter Papers, WHRC.

[28] Ernest LeNeve Foster to Edward Shovin, January 23, 1892, Foster to Messrs. Shovin and Lawyer, February 26, 1892, Foster lease with Shovin, March 15, 1892, all in Ernest LeNeve Foster Papers, DPL; Henry E. Wood, "I Remember," p. 34, Wood Papers, HL.

A group of Cornishmen would agree to work a mine or portion of it, to pay their own expenses, and to reimburse the owners with a fixed percentage of the profits. These arrangements appealed to the owners because the Cousin Jacks asked for only 10 to 15 percent of gross receipts. The miners sacrificed higher rates in return for a guaranteed fixed percentage of the ore processed, and they prospered; one group reportedly earned $100,000 in twenty months of leasing.[29]

There were some variations in the royalty arrangements in leases. In the Arizona Territory, Henry Wickenburg leased out the Vulture Mine for a flat fee of $15 per ton of ore extracted. In these years, 1863–1865, the Vulture's ore returned about $80 per ton, so leasers who struck high-grade ore could accumulate a good grubstake. Tonopah's Tasker Oddie also leased his property for a fixed dollar amount on all ore mined; both Oddie and the leasers found this arrangement profitable. At nearby Diamondfield, J. C. Kinnear and Howard Cochrane negotiated a lucrative lease on the Daisy Mine. Henry C. Morris and some friends who leased the Fairfield vein in Fairfield, Nevada, raised $300,000 worth of ore in six months, though freight charges of $14 per ton reduced their earnings.[30]

As Morris and his colleagues discovered, expenses often cut into the gross receipts. If leasers had to haul and mill the ore as well as mine it, their costs rose accordingly. Additional costs imposed by dead work and workers' demands for regular shifts, though expected, made profits hard to predict. Leasers sometimes abrogated their agreements when their costs outstripped their expected earnings. After all, leasing was still mining. The risks remained; they were merely shifted from the owner to the leaser. For example, leasers, like owners, occasionally suffered from untimely labor strikes. The Leadville strike of 1896 stopped work on leases as well as owner-run mines.[31]

By 1904, one lease for properties near Silverton included an

[29] Perrigo, "Cornish Miners," pp. 93–94.

[30] Duane A. Smith, "The Vulture Mine: Arizona's Golden Mirage," *Arizona and the West* 14 (Autumn, 1972), 234; "Nevada," *Harper's Weekly* 47 (June 20, 1903), 1018; Mrs. Hugh Brown, *Lady in Boomtown: Miners and Manners on the Nevada Frontier*, pp. 36–37, 44; J. C. Kinnear, "Fifty Years in Mining, 1907–1957," pp. 23–24, J. C. Kinnear Papers, WHRC; Morris, *Desert Gold*, p. 24.

[31] John L. McNeil to Charles L. Hill, Leadville, March 16, 1897, Wolftone Mining Co. Papers, SHSC.

exemption from the continuous work provision in the event of strikes. The lease also contained a suspension clause operative in the event of "unusual mining accidents or casualties." Thus, by the turn of the century, some leasers were receiving limited guarantees that their leases would remain in force should they encounter unforeseen and unavoidable delays. But other hazards continued to plague leasing operations. One group of Georgetown leasers labored diligently on the Diamond Mine until, after having spent $100 on it, they discovered that it had previously been poorly worked and was extremely dangerous; they terminated their lease. In the early twentieth century, Robert Livermore discovered the dangers of leasing, when one property flooded and a second's vein pinched out.[32]

Meeting deadlines and making payments were necessary concomitants of leasing. Leasers assumed the obligations and liabilities of management, yet their resources were limited and their standing tenuous. Creditors demanded prompt payment, and the individual leasers rarely obtained the best terms from suppliers and local merchants. The leasing miners depended upon professional assayers and sent their ore to independent mills that could not be trusted. Henry E. Wood found that the millowners undervalued his shipments, increasing their profits at his expense. If the leasers paid milling or smelting charges at the lessors' plant, problems might prove even greater, because the owners could then keep tabs on the leaser as well as extract a little extra profit from the mining operations.[33]

Thus, though there were problems in each of the arrangements, whether miners selected wage earning, contracting, or leasing, they could earn a good income by nineteenth- and early twentieth-century standards. Their costs of living were high, but so too were the daily wages and the income opportunities provided by contracting and leasing, at least most of the time. How then does one explain the well-documented legacy of labor discord and individual grievances associated with mining? In part the problem can be traced to the dangerous

---

[32] Catron lease with Cooper and Bibb, December 15, 1904, Rennehan-Gilbert Papers, RCNM; Richard A. Pomeroy to L. C. Damarin, Georgetown, June 3, 1878, Stevens Mining Company Papers, DPL; Livermore, "Autobiography," pp. 85, 87–88, Livermore Papers, WHRC.

[33] Wood, "I Remember," pp. 25–26, Wood Papers, HL.

and debilitating character of mining. Beyond that, though, lay a conflict inherent in the relationship between the miners and their managers. Ultimately, miners' choices were circumscribed by managerial decisions over which they had little influence and no control.

7.

Miners and Their Managers

Of the sub-bosses, miners and smeltermen, I can truthfully state that I do not believe a finer body of men ever existed.[1]

These men were like a great many more we have had, were no good. . . . I find that in this part of the country it is almost impossible to get good men at all.[2]

BECAUSE the self-interests of managers and their miners diverged, disagreements and complaints naturally arose. Wages, hours, and conditions of labor troubled the miners, while managers worried about the dependability of their workers and how to control them on the job. High-grading was one of the most irritating problems management faced. As a former high-grader, Frank Crampton recalled that even preachers supported the universal miner's maxim "gold belongs to him wot finds it first." Thomas A. Rickard, writing in 1908, stated unequivocally that "rich ore invites theft. . . . In every case the sight of gold, or of minerals that are known to enclose it, provokes industrial kleptomania." Rickard concluded that high-grading cost Goldfield's mining companies a thousand dollars or more a day. He estimated that ore worth over $1.5 million was stolen from the three leading mines, the Mohawk, the Combination, and the Florence, and that only a fifth of that sum was recovered.[3]

Though there were scattered incidents of high-grading in silver mines, most ore theft occurred in gold mines for an obvious reason: only gold held such a high value that the theft of small quantities

[1] Arthur L. Walker, "Recollections of Early Day Mining in Arizona," *Arizona Historical Review* 6 (April, 1935), 35.

[2] J. C. Dickey to Irving T. Bush, Central City, October 16, 1897, Gregory Consolidated Mines Papers, Western History Collection, University of Colorado Libraries, Boulder (hereafter cited as CU).

[3] Frank A. Crampton, *Deep Enough: A Working Stiff in the Western Mine Camps*, p. 50; Thomas A. Rickard, "Goldfield, Nevada—IV: Rich Ore and Its Moral Effects," *Mining and Scientific Press* 96 (June 6, 1908), 774.

justified the risk. In Leadville's Ibex Mine, men expected to "make a few dollars" if they worked in one of the rich stopes. Even the ore awaiting shipment on the dumps proved fair game for avaricious miners. Lewin Barringer proposed that ore from his Arizona mine be immediately loaded and shipped rather than allowed to accumulate on the ore dump with "every opportunity given to Mexicans and other evil disposed persons to run away with choice samples."[4]

Facing constant temptation, men were thought fools if they did not avail themselves of the opportunity to high-grade. Theft reached epidemic proportions in Cripple Creek and the boom towns of Nevada. Frank Crampton pocketed enough money from his "work" to travel first to the West Coast and then back to New York for an extended visit with his family. Charlie Siringo, the Pinkerton cowboy, reported that one young Winnemucca, Nevada, miner cleared seven thousand dollars in a few months of high-grading at the National Mine.[5]

Initially, thieves made caches of their pockets, boots, lunch pails, and timbermen's sacks. However, as companies countered the rampant thievery, their employees devised more sophisticated paraphernalia. Some men fitted under their clothes specially designed harnesses, which could conceal and transport as much as seven pounds of ore. Special garments with hidden or enlarged pockets aided other miners; even tools and dinner pails were rigged with secret compartments and false bottoms to accommodate stolen ore. Siringo claimed that a Winnemucca high-grader strapped an eight-pound chunk of ore to his stomach. Though temporarily troubled by pains, the man escaped with fifteen hundred dollars' worth of ore. Thomas Rickard reported that bold souls occasionally planned and executed the theft of prepared sacks and even whole carloads of ore. Activities on this scale lay beyond the pale of "acceptable" high-grading, and owners went to great lengths to recover such losses.[6]

[4] Report of W. E. Giese's Leadville Operative, June 10–June 12 and July 20 [1899], John F. Campion Papers, CU; Lewin W. Barringer to Daniel M. Barringer, February 11, 1896, Daniel Moreau Barringer Papers, Western History Research Center, University of Wyoming, Laramie (hereafter cited as WHRC).

[5] Crampton, *Deep Enough*, pp. 47, 55; Charles A. Siringo, *Two Evil Isms: Pinkertonism and Anarchism*, p. 101.

[6] Siringo, *Two Isms*, pp. 101–102; Rickard, "Goldfield," p. 775; Robert

But management found it difficult to thwart ordinary high-grading because of the acquiescence, if not the outright complicity, of the shift bosses and other minor supervisory personnel, who overlooked the trangressions of friends, relatives, or fellow countrymen. Thus Cornishmen found the pickings safest when working under a Cornish shift boss; similarly, the Irish, Swedes, and Americans pocketed ore under the loose supervision of their fellow nationals. John F. Campion learned from an informant that his miners discussed the ease with which ore could be pirated from the Ibex. "Oh, it is no trouble for those fellows who work in the rich stopes to get all the rich ore they want. . . . Of course you know when the bosses are careless the men will take advantage of it and make the most of it." Marshall Sprague, historian of Cripple Creek, maintains further that the local Mine Owners Association was unable to halt high-grading because of the alliances between members of the Association and crooked assayers and sampling firms.[7]

Assayers were prominent figures on the mining frontier, and the more plentiful they grew, the more troublesome they became. While assaying firms were necessary, the presence of several hundred, as in Goldfield, signaled a thriving business in illicit ore processing. Assayers had the perfect opportunity to mill and smelt the high-grade ore; they operated independent offices, had access to all the means of reduction, and possessed the required skills. Rickard discussed one brazen fellow who reduced the ore to bullion and then sold his tailings to an accomplice who engaged in mine salting. Given these temptations, the honest assayer stood as a monument to virtue in camp society. When he became an assayer, Crampton declared that "I had done my share of high grading with the rest of the stiffs in Goldfield, but I drew the line at melting it into bars." Unfortunately for the managers, Crampton's type was offset by firms like William Shea & Company of Golden, Colorado, which proudly advertised, "Being supplied with the neces-

---

Livermore, "An Autobiography," p. 91, Robert Livermore Papers, WHRC; Frank P. Tondel, "As I Remember Goldfield," *Nevada Historical Society Quarterly* 3 (Summer, 1960), 16–17; Marshall Sprague, *Money Mountain: The Story of Cripple Creek Gold*, pp. 204–205.

[7] Reports of W. E. Giese's Leadville Operative, June 10–June 12 and June 14–June 18 [1899], Campion Papers, CU; Sprague, *Money Mountain*, p. 312, also p. 205.

sary apparatus, we are prepared to manufacture *natural* nuggets (of gold) to order on short notice."[8] Though this blatant advertisement of services was intended to secure customers and notoriety in the boom camps of 1860, its spirit lingered after the boom towns settled into established communities. In a society that often measured individual accomplishments by material acquisitions, ethics paled in the presence of greed.

Assayers were not the only recipients of stolen ore; saloon-keepers, gamblers, jewelers, and local merchants also purchased high-grade. Some merely accepted it as payment for debts and services rendered, but less scrupulous persons worked as professional fences. Jewelers and gamblers often trafficked in high-grade because they had numerous opportunities to unload the pilfered ore. The temptation to accept or deal in high-grade was great, since customarily only 50 percent of the appraised or refined value of the stolen ore went to the miner. The remainder covered the costs and risks of the dealers. When the competition among the fences was stiff, dealers' returns fell to as low as 10 percent of the ore plus a fixed handling charge. However, even when their share of the take was high, miners frequently discovered that their treasures were undervalued or that they were otherwise defrauded. After all, the entire operation was grounded in theft and dishonesty.[9]

Management personnel devoted considerable time, energy, and thought to the problem of high-grading. Since there was widespread sympathy for high-graders, it proved difficult to apprehend and convict the culprits. One writer recently suggested that much of the public sympathy for high-graders stemmed from an idealized sensitivity to the plight of an individual confronting an impersonal corporation. In any event, employers had the utmost difficulty convincing juries of miners that high-grading was a punishable offense. As Rickard noted in 1908, "Even when the circumstantial evidence seems convincing,

[8] Rickard, "Goldfield," p. 776; Crampton, *Deep Enough*, p. 60; "Nuggets Made to Order," advertisement, *Denver Rocky Mountain News*, September 15, 1860.

[9] L. C. Kinikin, "Early Days in Telluride," *Colorado Magazine* 26 (January, 1949), 24–25; report of Operative R. C., June 29 and June 30 [1899] and report of W. E. Giese's Leadville Operative, July 1 [1899], Campion Papers, CU.

the local justices are afraid to convict; public opinion regards 'high-grading' as rather humorous, the labor unions preach a doctrine by virtue of which a little rich ore is the miner's perquisite, the rush to become wealthy possesses everyone, and in the excitement the moral law is left high and dry."[10] Yet management continued to press charges against suspects, and owners like Daniel M. Barringer found occasional moments of satisfaction when the legal system brought offenders to justice. "I am glad to see that you have bound over the two Cornishmen to the Grand Jury of Pima County in $1250 bond each, and that you are hopeful that their lawyer will not be prompt to get them transferred to Cochise County." Later Barringer was delighted to learn that "the thieving Cornishmen will have to remain in jail at Tucson until the April Term of Court." Other managers noted with satisfaction that Charles Anderson of Cripple Creek had been jailed under a Colorado statute that made high-grading grand larceny. Larceny was always hard to prove, however, because prosecutors had to verify the origins of the ore and catch the miner stealing before they could obtain a conviction.[11]

Usually unable to secure relief or justice through normal legal channels, capitalists took matters into their own hands. For example, in Cripple Creek, mineowners drove assayers, fences, and notorious high-graders from the district. Elsewhere, owners increased the supervision and surveillance of their employees, dismissing and blacklisting ore thieves. In the twentieth century the change house became the most effective means of combating robbery. Closely scrutinized as they went on and off their shifts, miners had to wear special work clothes that were changed at the shift's end. Some managers required all men to strip and shower before they could leave the company premises. These procedures made it increasingly difficult to high-

[10] Walker De Marquis Wyman, Jr., "The Underground Miner, 1860–1910: Labor and Industrial Change in the Northern Rockies" (Ph.D. dissertation, University of Washington, 1971), pp. 127–128. Wyman's industrial analysis of high-grading does not explain why Forty-niners, Fifty-niners and later boomers rationalized the practice and stole ore. Simple greed, widespread complicity, and the idea that gold belonged to its discoverer underlay the toleration accorded high-grading. Rickard's argument is found in "Goldfield," p. 775.

[11] D. M. Barringer to John Brockman, November 21 and December 27, 1901, Barringer Papers, WHRC; "Denver, Colorado," *Mining and Scientific Press* 96 (1908), 775.

grade. In Goldfield, change houses proved so effective that their elimination became a central issue in the strikes of 1907 and 1908.[12]

Some owners hired informers and private investigators as an alternative or adjunct to change houses and legal prosecution. Cripple Creek's James T. Burns explained that district's use of detectives to David H. Moffat: "We have had detectives working in the district for the last three months in this work, and we have succeeded in arresting quite a number of ore thiefs among whom are the notorious Vaneck Bros, and we expect to catch another batch this week and expect ere long to remove the business of ore stealing from the camp for good." In Leadville, John F. Campion employed several investigators to report regularly on ore thieves and their accomplices. His operatives related everything from braggarts' claims and rumors of high-grading to occasional eyewitness accounts of stealing or fencing. When the evidence against a miner seemed conclusive, Campion dismissed the offender and told other employers about the culprit's activities.[13]

Managers often claimed that their employees were lazy and careless. The owners bemoaned miners' wastefulness and their theft of mining supplies. Employers viewed unions with increased alarm after 1893, when miners from Colorado, Idaho, Montana, South Dakota, and Utah met in Butte to form the Western Federation of Miners. Precipitated by the labor conflict at Coeur d' Alene, the impulse for unionization was linked to the Panic of 1893, the technological and organizational changes in the mining industry, and the need for regional unity among the West's union miners. In response, the entrepreneurs organized themselves and resorted to extralegal espionage as the most effective way of identifying and eliminating their foes. Fearing a working-class conspiracy, the capitalists hired spies and detectives to entrap or provoke their own employees. The mutual mistrust consequently intensified and often prompted civil strife.

The informants were also used against other activities as well.

<hr>

[12] Sprague, *Money Mountain*, p. 312; Kinikin, "Telluride," pp. 24–25; Tondel, "Goldfield," pp. 16–17; Crampton, *Deep Enough*, p. 50.

[13] James T. Burns to D. H. Moffat, Cripple Creek, June 18, 1897, Eben Smith Papers, Western History Department, Denver Public Library (hereafter cited as DPL); reports of W. E. Giese's Leadville Operative, May–July [1899], report of "P. B." [C. L. Knuckey], June 20, 1900, Campion Papers, CU.

According to Leo J. Keena, many laborers turned to prospecting in their spare time and picked up company fuse, blasting caps, and a stick or two of dynamite for personal use. They regarded this petty larceny as a "loan backed by the best security in the world—the jack pot." The mining executives, observing all too many "loans" and all too few "jack pots," tried to exercise greater control over the distribution, allocation, and use of mining supplies. John Campion's informers and agents recovered missing tools, located caches of powder and caps, and identified idlers and malcontents. His operative "C. S." even reported the rumor that several miners had tied two cats' tails together and hung the animals from a nail for a cat fight. Such reports confirmed Campion's distrust of his miners.[14]

In Campion's eyes, union membership invested an individual with all the undesirable attributes linked to laboring miners in general. If a miner belonged to the union, thought Campion, then he was careless, lazy, dishonest, intemperate, rowdy, incorrigible, and a socialist-agitator. This simplistic stereotype made no allowances for the actual capabilities and faults of individual union miners, but it did reinforce the owners' fears, brought on by the real threats unions posed to the freedom of action and fiscal policies of mining administrators and corporations. Under optimum conditions a strong union might actually circumscribe the mineowners' choices; for example, a union might obtain a standard wage, limit the introduction of new technology, or define the pace or scope of worker's tasks. Nonetheless, unions rarely attained such power, though the unique example of the Comstock, with its four-dollar-per-day wage, was not lost on mine managers.

Individually or collectively in mineowners' associations, employers retained agents charged with surveillance and sabotage. The migratory and multinational character of the mining labor force made it relatively easy for men like Charlie Siringo to infiltrate local unions. Campion planted both needy local men and agents of the Thiel Detective Agency as spies. For several years after the Leadville strike of

14 Leo J. Keena, "Cripple Creek in 1900," pp. 4–5, Leo J. Keena Papers, State Historical Society of Colorado, Denver (hereafter cited as SHSC); reports of Operative C. S., June 8, 9, 14, 15, 16, and 17, 1897, report of Operative R. C., July 21, 1897, and reports of P. B., May 26 and June 3, 1901, Campion Papers, CU.

1896 Campion received regular reports on the Cloud City Miners'
Union from C. L. Knuckey, a miner and sometime union officer.
Knuckey, alias P. B. in his reports, was a cripple who needed steady
employment and the extra twenty-five dollars per month that collabo-
ration earned. Though P. B.'s reports express sympathy for the
owners and especially for Campion, the evidence suggests that money,
not principle, was his prime motivation.[15]

Some spying assignments were more specific than others. There
was no doubt about the mission Campion assigned one Thiel operative in 1898. "I wish you would have your operative here find out
who the parties are who have contributed $600 which it is said has
been raised by the Miners Union to pay the expenses of its [Labor
Day] picnic on the 14th [of August]." Two years later, S. W. Mudd,
manager of Campion's Leadville operations, wrote to W. E. Giese
asking him if he could "furnish . . . an Austrian miner for work in and
about Leadville," a request linked to the efforts of the Western Federation of Miners (WFM) and its Leadville local to organize immigrants. From Arizona, the Burns Detective Agency received a 1916
request "for one man who is a first class miner and good machine
man to work in my mines here among our other men. The purpose
would be to keep me advised as to any agitator I might have in my
employ."[16]

Informants received a fixed fee or fees in addition to the wages
earned as miners. C. L. Knuckey's standard twenty-five-dollar fee
was increased when he ran unusual risks or when his work required
peculiar expenses. S. W. Mudd reported that a Thiel agent was paid
thirty-three dollars for work performed in September, 1899. Pinkerton's Charlie Siringo later reported that "my salary was raised several
times during my stay with the agency and my overcharges against
clients shot up more than my salary." Operatives were paid either by
the individual managers or by assessments on the members of mine-

[15] C. L. Knuckey's activities and concerns can be traced through his
correspondence with S. W. Mudd and his reports under the alias P. B., in
Campion Papers, CU.

[16] J. F. C. to G. H. Thiel, August 20, 1896, Campion to W. E. Giese,
August 6, 1898, S. W. Mudd to W. E. Giese, November 1, 1900, all in Campion
Papers, CU; unsigned letter [Celora Stoddard] to Burns Detective Agency, No-
vember 28, 1916, Stoddard Family Papers, Special Collections Department,
University of Arizona Library, Tucson (hereafter cited as UA).

owners' associations. During the Leadville strike, the association established one levy at "3 and ½ per cent on the aggregate of the pay rolls for May and June," a sum that covered the costs of combating the strike as well as of retaining private detectives.[17]

Charged with locating all radicals, agitators, or anarchists, agents infiltrated the local unions and identified malcontents and union leaders. They reported the actions and movements of the union leaders, betrayed supervisory personnel who were sympathetic to the union, stole secret proceedings and minutes of meetings, and in other ways revealed union members and practices to employers. Though both the WFM and its successor, the International Union of Mine, Mill and Smelter Workers, were aware of this infiltration, the industrial union ideal of openness and the mobility of the mining work force precluded the effective screening of men seeking membership. As early as 1903, President Ed Boyce reported that a spy serving as recording secretary of the Colorado City local had reported all new members to the United States Reduction and Refining Company, which promptly dismissed them. As Ben Goggin viewed the Arizona labor troubles of 1916 and 1917, very little had changed. Once again union men faced dangers from spies, though in 1917 the informants were placed primarily in the WFM's rival, the Industrial Workers of the World (IWW). Modern historians have generally confirmed the judgments of contemporary union leaders. The scholars have uncovered extensive evidence of surveillance and even the occasional appearance of *agents provocateurs*. Inciting discord to heighten the public reaction against unions, these professional agitators did much to undermine the aspirations of thousands of working miners.[18]

[17] A. B. [S. W. Mudd] to C. L. Knuckey, December 30, 1899, and Mudd to H. I. Higgins, October 11, 1899, both in Campion Papers, CU; Siringo, *Two Isms*, pp. 8–9; Charles T. Carnahan to John F. Campion, Leadville, September 19, 1896, Campion Papers, CU.

[18] *Official Proceedings of the Eleventh Annual Convention of the Western Federation of Miners of America, Denver, Colorado, May 25–June 10, 1903* (Denver: Western Newspaper Union, 1903), pp. 26–29; "A Gila Monster," *Miners' Magazine* 18 (April, 1917), 3; Richard H. Peterson, "Conflict and Consensus: Labor Relations in Western Mining," *Journal of the West* 12 (January, 1973), 7–8; Philip Taft, "The Bisbee Deportation," *Labor History* 13 (Winter, 1972), 6, 8–9; James Ward Byrkit, "Life and Labor in Arizona, 1901–1921: With Particular Reference to the Deportations of 1917," (Ph.D. dissertation, Claremont Graduate School, 1972).

Confronting an unpredictable national government and seemingly uncontrollable external threats from the economy, mineowners and their associations concentrated on stabilizing those variables they could regulate. By applying improved technology and systematic management, owners could reduce the inherent uncertainties of mining. The other component of the operation subject to control was the labor force. Managers believed that wages should be adjusted according to economic fluctuations and that miners' wages should be linked to the prevailing cost of living. However, it is important to remember that management consistently opposed wage increases that would have kept pace with the inflationary trend before and during the American involvement in World War I. The preference for wage reduction inevitably brought management into conflict with the aspirations of their employees and the WFM.[19]

Mineowners experimented with several methods of wage reduction. One simple approach was to bring more men to a camp than were needed for the work available. Such a labor surplus increased the competition for jobs and lowered the pay level job-hungry workers were willing to accept. The *Virginia City Enterprise* described such a situation: "There are miners enough and workmen enough of every kind belonging here on the Comstock lode to do twice the work there is at present to be done in any branch of business. The sooner this is understood abroad the better." While there is no evidence that Virginia City's unemployment had been artificially created, twentieth-century miners believed that they did see attempts by employers elsewhere to create a pool of unemployed workers. In 1916 Charles H. Tanner, miner, wrote to the *Miners' Magazine* and denounced the Arizona copper companies that paid high wages to attract men from other locales. Then, according to Tanner, the firms selected "efficiency men" whose work rates were used as production standards applied to all employees. Workers unable to maintain the pace were discharged and replaced by others. A month after Tanner's charge, G. E. Francis and Steve Oberto, Leadville miners, challenged Philip Argall's assertion that miners there earned $4.00 to $4.50 daily; instead, they reported wages ranging from $3.30 to $3.85 per day. Francis and

[19] Vernon H. Jensen, *Heritage of Conflict: Labor Relations in the Non-Ferrous Metal Industry up to 1930*, pp. 88–92; Melvyn Dubofsky, *We Shall Be All: A History of the Industrial Workers of the World*, p. 27.

Oberto maintained that Argall's misrepresentation was a deliberate attempt to attract unemployed men, whose presence would strengthen local employers in their struggle against the union.[20]

Often employers took a cooperative approach to the wage question, knowing that they could "starve" workers into submission. In 1864, Comstock owners and operators coordinated their efforts to reduce wages. Colorado executives pledged to support wage reductions during the Cripple Creek crisis of 1894. By a formal declaration they agreed "to post [the] 9 hour day at $3.00 in their mines at 12:00 a.m. on Jan. 20, 1894. The new rate and hour scale to become effective Feb. 1, 1894. The undersigned owners agree to aid one another and refuse to negotiate with the union, but only with the men individually." Similar agreements proved successful in Leadville in 1896. In 1907 Kenneth L. Fahnestock reported discussions among Leadville's operators for a joint "reduction in the scale of wages for the high priced men," that is, the most skilled employees: engineers, firemen, cagers, blacksmiths, and timbermen. As Fahnestock explained, "The idea is not to pay anyone *less* than three dollars, but to cut those getting $3.50 and over, fifty cents per day."[21]

Not relying exclusively on cooperative tactics, individual owners on occasion acted by themselves, although independent measures subjected companies to greater and more effective union pressure. For example, the WFM successfully resisted an attempt by the manager of the Sun and Moon mines of Idaho Springs, Colorado, to reduce the wages of machine-miners' helpers. Under union pressure the company rescinded its reductions and abandoned a plan to import strikebreakers from Joplin, Missouri.[22]

As the WFM emerged as a force with which employers had to reckon, mineowners became increasingly concerned with identifying

[20] Anthony Hamilton to Frank [Dexter], Virginia City, June 6, 1877, January, 1878, and March 31, 1878, with clipping from *Virginia City Enterprise*, [Gideon] Anthony Hamilton Papers, Huntington Library, San Marino, California (hereafter cited as HL); Charles H. Tanner, "Conditions in the Mining Camps of Arizona," *Miners' Magazine* 17 (May, 1916), 6; "Communications," *Miners' Magazine* 17 (July, 1916), 8.

[21] Agreement among Cripple Creek Mine Owners, [January, 1894], James John Hagerman Papers, WHRC; Merrill Hough, "Leadville and the Western Federation of Miners," *Colorado Magazine* 49 (Winter, 1972), 19–34; Fahnestock to John F. Campion, October 28, 1907, Campion Papers, CU.

[22] *Proceedings of WFM, 1903*, pp. 117–118.

and eliminating union members from the labor force. One of the most effective means of assuring a "reliable" (nonunion) work force was the issuance of "rustling cards," which employers distributed to men they believed were "all right." The cards allowed employers to keep close tabs on the resident miners; if a man did not have a card, he was effectively blacklisted. Operators not only required compliance with antiunion practices among their own employees but also required that leasers and contractors implement their policies. Letters of reference and inquiries directed to former employers facilitated decisions about the reliability of recently arrived miners. For example, in letters of reference for miners leaving the Leadville area, Fahnestock used terms like "over in our yard" and "all right in every respect" to assure prospective employers that the man either had not belonged to the union or had totally renounced any prior affiliation with it. These procedures increased the cooperation and communication among employers and multiplied the difficulties for union miners. Men like Telluride's Vincent St. John, noted labor activist, found it impossible to secure employment under their own names. Union activity marked a miner as a malcontent and threatened his ability to pursue his chosen vocation.[23]

Managers and their miners found themselves at odds over more than high-grading, indolence, theft, and unions. As the owner placed his profits in the laborer's hands, so the miner entrusted his life to his employer. Miners' safety and the duration of their employment often depended upon whether operators conducted a thorough and prolonged exploitation of the property or whether they "gutted" the works in order to reap quick profits. Careful development ordinarily meant that the crosscuts, stopes, tunnels, and shafts received regular inspections and maintenance. On the other hand, rapacious capitalists and their managers gave scant attention to careful practices and pro-

[23] A. C. Van Cott to Messrs. Gallagher and Duffay, Cripple Creek, October 20, 1904, R. J. Ryan to Daniel McCarty, Cripple Creek, September 14, 1905, Mary McKinney Mining Company Papers, CU; Fahnestock to Generals John Chase and Sherman Bell, Leadville, September 3, 1903, Fahnestock to A. E. Carleton, February 3, 1905, Fahnestock to Secy., Cripple Creek Mining Assoc., April 18, 1907, Fahnestock to A. L. Arnold, January 16, February 5, 11, 27, and 28, March 12 and 24, April 25 and 29, June 16, 23, and 30, and July 10 and 30, 1908, Campion Papers, CU; Jensen, *Heritage of Conflict*, pp. 260–261, 357–360, 381–410.

cedures; they sought immediate profits and cared little about the long-term development of the mining properties.

Daniel M. Barringer, a mining engineer and investor, operated his mining ventures in a rational manner that included the careful working of existing ore bodies, additional exploratory work where it seemed justified, and the acquisition of adjoining or promising prospects. However, Barringer and his agents encountered constant resistance from minority stockholders who, concerned about the size and regularity of corporate dividends, held sound mining practices to be of decidedly secondary importance. In 1902 Barringer candidly told Sam W. Cheyney, "We have control of our company absolutely in our hands and so long as we are managing it ably and honestly and crowding the work as rapidly as is possible under the circumstances, they have no right whatever to make any complaint." However, by 1910 Barringer's agent explained to him that he "found it absolutely necessary" to pay the size dividend the stockholders demanded.[24] When, as in Barringer's case, operations were partially circumscribed by the demands of shareholders, the profit motive and the nature of the corporate structure became a factor in mining practices.

Within the bounds allowed by stockholders, management decided which of the possible ore removal techniques they would employ. Writing in 1929 E. D. Gardner and D. J. Parker clearly explained the link between mining practices and the individual hard-rock miner. "The responsibility for choice of the mining method rests with management; moreover, responsibility for accidents in any mine rest ultimately upon the same shoulders." That competent supervisory personnel were invaluable in mining operations had been impressed upon young E. D. Gardner as early as 1912.[25]

The individual laborer was caught between the demands of quality and quantity, careful development and "gutting," long-term gains and immediate profits. He had no control over the environment and pressures under which he would work; often his only choice was

[24] Daniel M. Barringer to Sam W. Cheyney, June 7, 1902, Smith to Barringer, June 3, 1910, Barringer Papers, WHRC.

[25] E. D. Gardner and D. J. Parker, "Safety Organizations in Arizona Copper Mines," U.S. Bureau of Mines, *Technical Papers* 452 (1929), 2; Joe [Manwaring] to E. D. Gardner, Bingham Canyon, Utah, January 24, 1912, E. D. Gardner Papers, WHRC.

between working and starving. In 1908 Frank Crampton discovered exactly how carelessly a mine could be managed. Speaking of his experiences in an old Utah mine he noted: "On that job I learned the most important lessons of my life: how not to operate a mine, and how not to treat or handle men. . . . The mine was operated in the old tradition that nothing but dollars was of importance."[26]

As Crampton's statement suggests, miners endured the callousness and personal pettiness of some operators. For example, Duane Smith argues that the Leadville strike of 1880 stemmed jointly from a policy in the Chrysolite Mine that forbade all unnecessary conversation underground and from an unexplained work stoppage at the Little Chief Mine. Smith argues further that the crisis may have been carefully fabricated by certain of Leadville's mineowners and promoters, who hoped to camouflage mismanagement of the Chrysolite Mine. Unable to restore confidence in the flagging Chrysolite, they were nonetheless able to postpone payment of one dividend and perhaps to sell some of their own Chrysolite stock before their malfeasance was discovered. Elsewhere, men found themselves deprived of their legitimate wages and learned to be "suspicious of companies that do not make payments in cash as they go." Deliberate abuse of miners' rights demanded legal action, but individually and collectively miners often lacked the resources to challenge their employers. Even liens on property and materials were often useless, especially since the concerns most likely to default were the prime candidates for bankruptcy.[27]

Men who found work in remote mines or in company towns encountered additional management supervision. When sleeping in a company dwelling, eating company "grub," or trading at the company store, employees increased their dependence. Although the quality of life in company towns could be better than elsewhere, even management's good intentions might seem tyrannical, given the degree of the miners' subjection.[28] In company camps or elsewhere, the routine

[26] Crampton, *Deep Enough*, pp. 101–103.

[27] Duane A. Smith, "Colorado's Urban-Mining Safety Valve," *Colorado Magazine* 48 (Fall, 1971), 314–315; idem, *Horace Tabor: His Life and the Legend*, pp. 134–148; F. A. Prescott to Robert T. Polk, Austin, Nevada, August 9 and September 20, 1864, Henry Douglas Bacon Papers, HL.

[28] William D. Haywood, *Bill Haywood's Book: The Autobiography of William D. Haywood*, pp. 22, 57.

interaction of miners and managers could take on other, subtle over-tones; managers, for example, encouraged and exploited prejudices among workers.

In 1883 Eliot Lord observed that Comstock officials played on traditional national rivalries. By mixing work forces, managers thwarted national cliques' connivance to loaf. In the fluid society of the West, traditional allegiances were links with a pre-Western past. Thus German miners associated with other Germans, Irish with Irish, Cornish with Cornish, and so forth. Managers often tried to use these links and exploit animosities to strengthen their own position in deal-ing with employees. In Leadville, Campion's private investigators reported conflicts between the Irish and the Cornish and between Americans and immigrants. While recognizing the importance of co-operation, the Irish union leadership remained suspicious of Cornish-men, Welshmen, and native-born Americans. The capitalists en-couraged this enmity, which hampered the organization of Leadville's miners. Elsewhere, Charlie Siringo witnessed the manipulation of men and the destruction of unions by the careful "mixing of nationalities" and the cultivation of ill will among various immigrant groups.[29]

Not all managerial preference for particular nationalities was intended to foment national rivalries, however. For example, early Nevada and Colorado managers generally preferred Cornish workers, because they were experienced miners. By the end of the century, company officials favored Italians and Eastern Europeans because they accepted lower wages. (Much the same rationale applied in mines that recruited Missouri lead miners and Appalachian coal miners.) According to Albert H. Fay, these immigrants came from rural European backgrounds and had little or no prior experience in min-ing. Since many of them were illiterate, they accepted and tolerated lower wages and poorer working conditions than American or British miners would accept. In addition, they held the most hazardous oc-cupations underground. Eighty-two percent of them worked as pick miners, trackmen, car runners, machine runners and helpers, and

[29] Eliot Lord, *Comstock Mining and Miners*, p. 382; D. W. Brunton to J. F. Campion, Aspen, August 24, 1895, reports of Giese's Leadville Opera-tives, August 6, 8, 10, and 21, 1898, and April 1, 2, 3, and 7 and October 15, 1899, Campion Papers, CU; Siringo, *Two Isms*, p. 109.

loaders, all dangerous jobs. By contrast only 61 percent of the English-speaking miners held these positions.[30]

Throughout the era, miners who accepted lower wages angered their colleagues, native and immigrant alike. Ironically, low-paid workers often did not know that their wages were below those paid others. The newly arrived immigrant needed a job in order to support himself and perhaps a family. Facing starvation and a lack of shelter, immigrants accepted jobs when and where they could find them. Helen Zeese Papanikolas chronicles the plight of many Greek immigrants who found positions with the help of Leonidas G. Skliris and his three brothers. Employed as a labor agent by the Utah Copper Company, the Western Pacific Railroad, the Denver and Rio Grande Railroad, and coal operators in Wyoming's Carbon County, Skliris exploited Greek immigrants in Idaho, Wyoming, Colorado, Nevada, California, and Utah. Active as the intermediary between American corporations and his countrymen, Skliris tyrannized Greek immigrants and built a fortune by collecting employment fees when the newcomers received jobs. His greed led to unfortunate consequences in 1912 during the bitter strike by Bingham Canyon's miners against the Utah Copper Company.[31]

Then as now, Utah Copper's open-pit mine in Bingham Canyon was one of the largest in the country. Utah Copper employed a large force of the so-called new immigrants, principally Greeks, Austrians, and Italians. Their wages were low, and they lived in company-owned quarters. Not only had the Greeks been recruited by Skliris, but they had also been pressured into patronizing his store and paying periodic assessments to retain their jobs. While legitimate anger at the grasping, graft-tainted labor agent precipitated the strike, the Bingham Canyon trouble was widely reported to be an immigrant insurrection in newspapers that used value-laden terms like "foreign army," "anarchists," "aliens," and "foreign element" (as opposed to a "white

[30] Albert H. Fay, comp., "Metal-Mine Accidents in the United States during the Calendar Year 1918," U.S. Bureau of Mines, *Technical Papers* 252 (1920), 96.

[31] Helen Zeese Papanikolas, "Toil and Rage in a New Land," *Utah Historical Quarterly* 38 (Spring, 1970), 115–116. See also newspaper clippings in Strike Scrapbooks, Daniel Cowan Jackling Papers, Bancroft Library of University of California, Berkeley (hereafter cited as BL).

element"). In this nativist atmosphere, the strikers and their grievances were dismissed as being radical and un-American. Only the
radical *New York Call* and the prolabor *Ogden Standard* denounced
the activities of labor agent Skliris and the policies of Utah Copper.
"In Bingham canyon, the Utah Copper Company, seeking the cheap-
est labor obtainable, has employed Greeks, Austrians, and other
foreigners, most of them new-comers who lend themselves readily to
the artful schemes of those who resort to the padrone system of
mulcting labor." Though as sensational as the antiworker accounts,
this report in the *Ogden Standard* rightly identified the paternalistic
labor methods employed by Skliris and the Utah Copper Company.
The strike itself dramatized Skliris's activities and forced his resigna-
tion. Even though other issues remained unresolved and the strike
languished, Greek miners celebrated Skliris's dismissal.[32]

Sharing the racial attitudes of their times, managers often cared
little about the plight of the immigrant miners. According to Frank
Crampton, some companies regarded all miners as "cheaper than
timber," and no miner was lower in status than an Eastern European
or Mexican. Characteristically, John F. Campion reproached his
brother George for hiring "Irish and Dago" miners at the Reindeer
Mine. "Some months ago I had an understanding with you that you
would hire these men yourself, but if this is the kind of stuff you have
been hiring, which I very much doubt, it don't look good to me."
Similar prejudices moved Lewin W. Barringer to caution his brother
about the "Irish Mexican greasers" in the work force around Wilcox,
Arizona. In 1921 Robert E. Donovan of the Old Dominion Copper
Company of Globe, Arizona, assured E. D. Gardner that aliens (Mexican-Americans), whether single or married, would be laid off before
any white men. In company towns, strict geographical segregation
isolated the Mexican-American community from its non-Hispanic
counterpart. In the 1860's one American manager labeled his workers
as "white men" and "Mexicans" or "Peons." Whites held the supervisory positions and were paid higher wages. It took little effort for
mineowners to foster racial discrimination so as to confound workers'
attempts at unification. Though the policy was temporarily expedient

[32] "Pass Prosperity to Bingham Strikers," *Ogden Standard*, September 21,
1912, Strike Scrapbook 1, Jackling Papers, BL.

for the owners, it left bitter legacies for those who had been oppressed.[33]

Not only immigrants but all miners were likely to suffer when their aspirations and welfare clashed with those of their employers. The miners sometimes reacted by forming associations, which are considered in the next chapter, but even as individuals, miners protested injustices. They clashed with management over such issues as accident liability and wage obligations. In the highly speculative mining industry, money was normally in short supply, and without adequate returns from their mines, even some of the strongest corporations defaulted, leaving workers unrecompensed. For example, Leadville's William H. Stevens was nearly lynched because he lacked sufficient funds to meet a midwinter payroll. Only the timely discovery that funds were available in Denver saved Stevens from his angry workers. In Kimberly, Utah, a mining company could meet its payroll only with scrip that it pledged to redeem. When the company went bankrupt and closed its mines, not only the miners but also the local merchants were left with worthless scrip. In another incident, in January, 1878, when Cecil C. Morgan assumed control of Georgetown's Little Terrible Mine, the company owed its wage and contract miners $22,697.00 for November and December. Though Morgan assumed responsibility for securing these back wages, the payroll continued to lag two months in arrears throughout that spring. Five years earlier Georgetown's St. Lawrence Mining Company had defaulted when W. Frank Witherell deserted his commitments and his employees. One angry contractor wrote to him demanding payment of his debt. "The reason of my sending you these few lines is wanting to know about the moneys that is due to me on the St. Lawrence Mining Company for Contract on the Silver Wing Lode and at the Crosscut Tunnel[.] I had against the firm ($3495.28) Thirty Four Thousand and Ninety Five Dollars & twenty eight cents [*sic*] and out of the above I received

[33] Crampton, *Deep Enough*, pp. 101–103; John F. Campion to George Campion, Leadville, 1905, Campion Papers, CU; Lewin W. Barringer to D. M. Barringer, February 11, 1896, Barringer Papers, WHRC; Robert E. Donovan to E. D. Gardner, March 29, 1921, Gardner Papers, WHRC; "Second Annual Report of the Ray Consolidated Copper Company," September 27, 1910, inset picture with separate townsites for English- and Spanish-speaking employees, Mining File, WHRC; Thomas M. Yenkes to H. D. Bacon, Pima County, Arizona, April 15 and June 16, 1865, and March 19, 1866, Bacon Papers, HL.

Three Hundred Dollars." Another creditor explained, "It is not necessary for me to say that I am *very needy*, or if you please, hungry for money." Elsewhere, miners encountered similar problems with delinquent payrolls and bankrupt corporations. Cornishman Richard Harvey complained bitterly to William Downing about delinquent wages Downing owed him. Harvey noted that the money "would be gratefully and needfully received," especially inasmuch as "it had been over 3 years" since the work had been performed.[34]

Few corporate agents were as concerned about their workers as James McGrath of Holmes, Wyoming, who was reluctant to employ men when he believed they might not be paid for their labor. In 1906 he told H. V. S. Groesbeck of Laramie, "I cannot see why some law cannot be enacted by our legislative body, and enforced for the protection of laboring men against snide companies—or so called Co hiring a lot of men to work, knowing well at the same time they have not the means to pay those men for their labor, thereby literally robbing these men of their only capital." Nearly forty years earlier J. G. Mahany had personally advanced money to Central City miners, even though he did not possess any company funds at the time of the disbursement. Like McGrath, he took pains to ensure that they were or would be fairly compensated for their labor. However, many corporations and their agents were less scrupulous.[35]

Wage-related issues spawned much of the labor discord that plagued the hard-rock mining industry. For example, while employers like Georgetown's R. Orchard Old claimed that being paid monthly satisfied the workers, the WFM urged management to pay miners bimonthly and more promptly. This issue precipitated a labor strike at

[34] Iva Evans Morrison, "William H. Stevens," *Colorado Magazine* 21 (July, 1944), 126–128; Josephine Pace, "Kimberly as I Remember Her," *Utah Historical Quarterly* 35 (Spring, 1967), 120; Diary of Cecil C. Morgan, January 6, 8, and 10, February 2, and May 4, 1878, DPL; Charles Mathews to W. F. Witherell, Georgetown, November 11, 1873, D. W. Willey to Witherell, Georgetown, November 11, 1873, W. Frank Witherell Papers, SHSC; Harvey to Downing, Central City, March, 1875, in "Old Caribou and Central City: Glimpses from the Letter Book of Cornishman Richard Harvey," Agnes Wright Spring Papers, WHRC.

[35] James McGrath to H. V. S. Groesbeck, Holmes, Wyoming, March 4, 1906, H. V. S. Groesbeck Papers, WHRC; J. G. Mahany to H. D. Bacon, October 15, 1868, Henry Douglas Bacon Papers, HL.

the Jesse Knight Mine in Bingham Canyon in 1917. When they quit or were discharged from a job, miners expected immediate payment of back wages so that they could seek another position at once, but the men were largely at the mercy of their employers in such situations.[36]

In most instances where miners were injured or went unpaid because of corporate negligence and insensitivity, miners had only one recourse, suit in a state or territorial court of law. Unpaid wages usually constituted a first or primary lien against corporate property that belonged to the person or persons employing the miners. The Federal Bankruptcy Statute of 1898 gave first priority to " 'wages due to workmen, clerks, or servants,' if 'earned within three months before the date of the commencement of bankruptcy proceedings, not to exceed three hundred dollars per claimant.' " This seemed to guarantee individuals legal recourse and a primary lien against company property. The actual legal picture was often much more complex; first, labor liens applied only to holdings of the company or individuals directly employing miners. Thus, where miners worked for contractors or leasers, the company actually owning the property was not liable for the operators' activities. The complexities increased when a corporation or individual leased property to others with an option to buy. Such men often described themselves as the "new owners," and the average miner was in no position to ascertain the leasers' legal standing. The miner believed the operators were in fact the owners and could act in that capacity. However, option holders were technically equivalent to leasers, and while the courts held them personally liable, judges exempted the property in question from the provisions of the miner's lien.[37]

Nowhere was the technical and biased nature of the law more

[36] R. Orchard Old to Subcommission on Mining of U.S. Industrial Commission, Georgetown, April 22, 1899, R. Orchard Old Papers, SHSC; "Human Beings Are Cheap," *Miners' Magazine* 18 (April, 1917), 8; minutes of executive board, WFM, June 9, 1902, Mine, Mill Union Papers, CU; "Miners at Chloride Get a Raise in Wages Which Prevents a Strike," *Miners' Magazine* 18 (July, 1917), 6.

[37] 30 Stats. 544 (Act of July 1, 1898), as quoted in Lawrence M. Friedman, *A History of American Law*, p. 482; Griffin v. Hurley, 7 Arizona 399 (1901); Walter C. Hadley Company v. Cummings et al., 7 Arizona 258 (1901).

apparent than in a suit brought by John Harper against Arizona's Independence Development Company and its stockholders, Thomas Farrish, J. C. Goodwin, R. G. Goodwin, and G. J. Ridenour. The company and each individual mentioned had mining claims listed in his respective name, all supervised by the Independence Development Company and its resident official, none other than R. G. Goodwin. Between October, 1906, and May, 1908, the company employed sixty-one miners to work on its properties. During these months $9,519.28 in unpaid wages accumulated as a liability. Throughout this period R. G. Goodwin continued to assure the miners that they would receive the wages due them. When the workers had still not been paid for their labor after many months, Harper, one of the miners, initiated a suit. The Arizona Supreme Court upheld liens against the six claims registered in the name of the Independence Development Company, but it exempted all of the claims registered to individual shareholders in the company. Thus the individual stockholders escaped their reasonable obligations through legal manipulation. The miners had no way of deciphering the legal situation, and when hired, they had undoubtedly assumed that the Independence Development Company owned all of the properties on which they worked.[38]

In 1912 the Arizona legislature revised the statute governing liability under the miner's lien and extended this right of lien to all property regardless of the technical ownership. Thus owners became liable for the wages and debts of contractors, leasers, and option holders. However, once again the courts frustrated the obvious intent of the legislators. All contracts with leasers or option holders made prior to enactment of the statute were exempted from the provisions of the 1912 law, even though the actual labor by miners occurred after the date on which the new law became effective. The Arizona Supreme Court argued that if the provisions of the statute were applied, it would constitute a "retroactive application of law against prior rights and interests" and would thereby unnecessarily burden the original contract. This decision convinced miners that they had only marginal chances of success if they sought legal relief. Furthermore,

[38] Harper v. Independence Development Company et al., 13 Arizona 176 (1910).

the high costs of litigation discriminated against the working man. Unpaid miners faced an unpleasant choice: moving on or initiating an expensive and protracted suit whose outcome was at best uncertain.[39]

The legal difficulties miners encountered in lien actions were magnified when mineworkers resorted to legal channels to redress grievances linked to occupational injuries. According to Lawrence M. Friedman, a prominent legal historian, "the modern law of torts [personal-injury actions] must be laid at the door of the industrial revolution, whose machines had a marvelous capacity for smashing the human body." Friedman argues:

> Every legal system tries to redress harm done by one person to another. The industrial revolution added an appalling increase in dimension. Its machines produced injuries as well as profits and products. The profits were a tempting and logical fund out of which to pay the costs of the injured. Moreover, the industrial relationship was impersonal. . . . But precisely here (to the 19th-century mind) lay the danger. Lawsuits and damages might injure the health of precarious enterprise. . . . The careful, conscious aim of the judges was to limit damages to some moderate measure.[40]

Mining fit well this more general characterization of industry undergoing profound technological change. Between 1860 and 1920 mining became increasingly mechanized, and mining machinery mangled each new generation of hard-rock miners. When miners initiated tort actions, they were blaming corporations for discomfort and disfiguration; such claims collided head-on with the prevailing reading of tort law.

Four important doctrines shaped the courts' interpretations of personal injury suits. "Assumption of risk" proved to be one of the harshest rules. This common-law principle absolved employers from responsibility in all accidents that resulted from "ordinary risks" of their employees' occupations. The courts held that the employees understood their occupational hazards and therefore assumed the liability for common injuries. Along with the assumption of risk judges upheld the "fellow-servant rule." Under this doctrine, the courts ordinarily acquitted employers when other company employees contribut-

---

[39] Foltz v. Noon, 16 Arizona 410 (1915); Oceanic Gold Mining Company v. Steinfeld, 16 Arizona 571 (1915).

[40] Friedman, *History of American Law*, pp. 409–410.

ed to or caused a worker harm. The loose interpretation of this concept occasionally exempted corporations even when managers or engineers directly ordered miners to perform dangerous tasks that resulted in injuries. In effect, these legal dicta fixed primary responsibility for safety on the employees rather than the management.

Joining these doctrines was the catchall concept labeled "contributory negligence," which, as applied in tort cases, fixed the burden of proof on the plaintiff. If the miner had contributed in any manner whatsoever to his own injury, then the defendant, the mining company, was exonerated. In practice, courts assumed that miners were negligent unless they could prove they were not; in effect, the victim was compelled to prove his innocence beyond a shadow of a doubt. The cumulative effect of these practices was to strip the miners of their rights to just compensation before the law. One last doctrine even prevented the immediate families of miners killed in accidents from obtaining relief. Nineteenth-century jurists and lawyers adopted the common-law practice that held tort suits to be peculiarly personal damage suits. Torts could be initiated only by the injured, whose legal rights accompanied them to their graves. Thus the wives and children of deceased miners were excluded or disqualified in tort cases because the aggrieved person had died.[41]

As early as the 1890's, the courts began to reexamine the plight of the plaintiffs. In part, this reconsideration stemmed from a shift of concern, both public and judicial, from the protection of the corporation to the protection of individual workers. More importantly, political reformers turned their attention to laborers and enacted legislation to protect miners from specific abuses. These laws regulated working conditions and the hours of labor; they established state bureaus of mines and labor, which were expected to investigate working conditions. Though actual inspection and regulation were often spotty, the mere presence of the legislation opened a new avenue of appeal. Courts could then hold corporations accountable when company officials had ignored the requirements of safety legislation. Thus, "statu-

[41] Ibid., pp. 412–415; James Weinstein, "Big Business and the Origins of Workmen's Compensation," *Labor History* 8 (Spring, 1967), 158. Weinstein estimates that while almost 70 percent of all industrial accidents stemmed from either the nature of the work or employer negligence, a mere 15 percent of victims recovered damages in civil actions.

tory negligence" replaced common-law tenets and helped redress the balance between injured miners and the men who managed them.[42]

Between 1883 when Eliot Lord wrote about conditions on the Comstock and 1929 when E. D. Gardner and D. J. Parker issued their study, "Safety Organizations in Arizona Copper Mines," the social responsibility for accidents shifted from the individual who had been injured to the corporation for which he worked. In 1929 Gardner and Parker stated categorically that the costs of accidents fell to the owners and "should be charged against cost savings of doubtful and unsafe practices." The changes in their legal liability induced company officials to search for new ways to protect themselves. Management often obtained employers' liability insurance, which could provide either specific insurance to cover the costs of litigation and adverse settlements or a more general coverage to pay legal expenses or benefits for injured workers who waived their rights to sue. Most officials preferred the comprehensive protection the second type offered. By the 1890's employers' need for protection was obvious, as Ernest LeNeve Foster observed: "The tendency grows every year for the miners to demand damages from companies for injuries received whilst in their employ." John Campion learned that reputable insurance companies merited the somewhat higher premiums they demanded. Having been victimized by the American Casualty Insurance and Security Company, he subsequently examined all policies carefully and became and remained a loyal and satisfied client of the London Guarantee and Accident Company, Ltd. When evaluating one competing firm, he noted that it limited its "liability to two years instead of six" and retained the right of cancellation as well as the prerogative to "return to the holder paper of the concern instead of cash."[43]

---

[42] Strike v. Ray Consolidated Copper Company, 16 Arizona 237 (1914); Weinstein, "Workmen's Compensation," pp. 158–160; Friedman, *History of American Law*, pp. 418, 421.

[43] Lord, *Comstock Miners*, p. 220; Gardner and Parker, "Safety Organizations," p. 1; Foster to Lichtenstein, Central City, January 2, 1892, Ernest LeNeve Foster Papers, DPL; John F. Campion to George W. Trimble, Leadville, January 3, 1899, Campion Papers, CU. Campion's troubles with the American Casualty Insurance and Security Company can be followed in correspondence between James K. Brown and Milton Smith, attorneys, and Campion, 1892–1895, Campion Papers, CU.

Liability insurance protected the owners rather than the employees. Though employees ordinarily paid part of the premium, they discovered that there were exceptions to the coverage. For example, death from natural causes or debilitation due to miner's consumption were not usually covered. Some insurance policies specifically excluded injury or illness attributable to "bad air" or mine gas. Furthermore, legal waivers included in the policies eliminated any chance for additional compensation. While miners grumbled about such provisions, the insurance did offer minimal protection and constituted a temporary alternative to tort action. Ultimately, government, industry, and laborers turned to workmen's compensation as a more satisfactory response to the problems of industrial accidents.[44]

Rather than acquiring liability or compensation insurance, some owners supported local miners' hospitals. Some companies assumed responsibility for injured men on a case-by-case basis, guaranteeing treatment at a nearby hospital. Other companies operated their own employee-compensation programs. Arizona's Pride of West Mining and Milling Company collected premiums and provided a company doctor and a general fund from which hospitalization expenses were paid. Company-supported hospitals and doctors were most common in company towns, and miners often complained bitterly about the treatment and compensation they received from company health personnel. Company officials provided the only supervision of the medical facilities, and, since management had a vested interest in the arrangements, officials were often reluctant to pursue miners' criticisms.[45]

Miners and managers formulated opinions of one another, which ranged from deep admiration to virulent distrust and even hatred. As previously noted, Leadville's John F. Campion was so suspicious of his miners that he retained detectives to discover which men put in a

[44] Thomas F. Daley to Campion, Denver, April 4, 1896, Daly to Ibex Mining Company, Denver, March 22, 1902, Campion Papers, CU.

[45] R. E. Mantius to Paul O. Esterhazy, Georgetown, January 8 and March 15, 1881, William A. Hamill Papers, DPL; S. W. Mudd to Sister Superior, Sisters Hospital, Leadville, November 3, 1899, Campion Papers, CU; Jesse Scobey to Thomas F. Daly, Washington, Arizona, August 16, 1902, Arthur Wifley Papers, CU; Joseph D. Cannon, "The Grievances at Miami and Globe, Arizona," *Miners' Magazine* 18 (August, 1917), 5; Kain v. Arizona Copper Company Ltd., 14 Arizona 566 (1913).

full shift and which did not. Some employers believed miners to be spendthrifts and described them as careless workers. M. O'Gorman, manager of the Middlemarch Mine in Cochise County, Arizona, struggled along from day to day with his crew. Describing his night shift workers as "originally a poor lot, physically and mentally," he had to cajole and badger them to perform the necessary work. It proved an unpleasant task that filled his diary with complaints.[46]

On the other hand, men like Arthur L. Walker lavished compliments upon the men who worked for them. John Hays Hammond once described the miner as "a man of exceptional probity, and it is but tardy justice to recognize this virtue." Dorchester Mapes of Wyoming's Rambler Copper and Platinum Company extolled his "enthusiastic force of fifteen experienced men, capable of handling every part of the work and coping with every difficulty that may arise and all working in perfect harmony." Though Walker spoke after the fact and Hammond and Mapes were seeking to impress investors, their respect for their employees was genuine. Even Fahnestock and Campion gladly provided glowing recommendations for individuals who had renounced or continually opposed the WFM local in Leadville, and, undoubtedly, some of this good will was reciprocated. The catch in this instance, as in others, was that union membership *ipso facto* colored management's assessment of a miner.[47]

Most employers labeled union men as lawless, vengeful agitators. "The laboring man (as a class) is a damphool. The only way you can gain access to his intellect is through the medium of his stomach. I can see no way that will be efficacious except to starve him to it." Campion and his associates filled their correspondence with denunciations of organized miners in Colorado and elsewhere. They saw unions as a threat to property and community standards of law and order. They vowed to "discriminate against [their] active members" and coordinated efforts to destroy unions and their members. Few supervisory personnel echoed the sentiments of Theodore F. Van Wagenen, who

[46] Diary of M. O'Gorman, January 21, March 16 and 24, and December 4 and 6, 1901, UA.

[47] Walker, "Recollections," p. 35; John Hays Hammond, "Suggestions Regarding Mining Investments," address delivered before Finance Forum of New York, December 22, 1909, Mining File, WHRC; Dorchester Mapes to A. C. Dart, Wyoming, November 14, 1916, Goodrich Papers, WHRC.

applauded organizations working to secure better living conditions and opportunities for laboring miners. In the prestigious *Mining and Scientific Press*, Van Wagenen asserted that "there is no question of their right to combine for this end" and argued that unionization was inevitable.[48]

The miners, for their part, had mixed opinions of their employers. Many resented management's power over them in this era when managers had few constraints of conscience or of law. Companies claimed the exclusive right to raise or lower wages, shorten or lengthen shifts, alter contracts, and declare leases forfeit. Some few conscientious and lucky miners escaped this domination by being promoted into the supervisory ranks. Most, however, worked in the mines as long as they were physically able. Having to concentrate on keeping the boss contented, miners discovered that as individuals they could have little or no influence over the decisions made. Their responses therefore became cooperative and defensive; they organized and joined unions in order to counter managerial domination. Under union banners, they resisted wage reductions and sought a living wage for all underground workers. They battled for shorter hours and better working conditions. Miners discovered that unions could indeed play an important, if precarious, role in improving their lot.

[48] J. J. Blow to Father Dyes, Leadville, November 22, 1896, J. J. Blow Papers, CU; Arthur Winslow to Mine Operators of Colorado, April 28, 1902, Eben Smith to John F. Campion, Denver, September 22, 1896, Campion Papers, CU; W. E. Mattes to J. J. Hagerman, Pueblo, May 28, 1894, Hagerman Papers, WHRC; Theodore F. Van Wagenen, "The Rights of the Miner," *Mining and Scientific Press* 96 (1908), 669–673.

8.

The Need to Organize:
The Miner and His Union

We miners of the Western states obtained the eight hour day;
In every state we organized and still kept up the pay;
We did not go to Congress, as lately has been done,
But through our economic strength, the shorter day was won.[1]

WHILE miners unionized for a variety of reasons, the primary factor in their decision to form a union was economic. Wage and hour issues dominate the history of Western mining unions, as of the American labor movement as a whole. Although economic questions were prominent among his purposes, the miner also sought through his union to enrich his social, cultural, and even political environment.

Unions benefited from the distinctive sense of cooperation, interdependence, and group solidarity that emerged from the crucible of the mining experience. This fellowship surfaced in local unions' support of their sick or injured members and in their aid and comfort for bereaved relatives. Beginning with the early Comstock associations and continuing into the twentieth century, local unions provided members illness, injury, and death benefits. Since few miners could afford insurance, the union's assistance, which in Cripple Creek included nursing care and ten dollars per week, was important. Local members carried this concern to the Western Federation of Miners' (WFM) conventions, where they proposed insurance funds and miners' homes for the aged, infirm, and destitute. While these issues received continuous and spontaneous delegate support, they were repeatedly tabled because of economic or philosophical objections.[2]

[1] J. P. Holohan, "The Six-Hour Day," *Miners' Magazine* 18 (February, 1917), 6.

[2] Frank Aley, "The Deep, Black Stope," *Miners' Magazine* 1 (February, 1900), 13; Richard E. Lingenfelter, *The Hardrock Miners: A History of the Mining Labor Movement in the American West, 1863–1893*, pp. 38, 51–52,

The miners also expected their unions to sponsor social events. Union meetings provided an atmosphere of conviviality and comradeship in which miners could meet, swap stories, and brag about how they had outwitted the "old man." Politics, international affairs, industrial relations, and mundane chatter mingled at union gatherings, and afterward there was always a chance to join old friends for a nightcap. Conventions brought the most active miners together to establish Federation policy and to relax in Denver or Salt Lake City. The unions also sponsored the Labor Day outings described earlier, held Christmas parties, and participated in Fourth-of-July celebrations. Established unions prided themselves on their union halls, which often included sitting or reading rooms. In 1902, in recognition of the broad social scope it had attained, the WFM even considered forming a women's auxiliary.[3]

So far, the common purposes of union members have been stressed; however, this emphasis glosses over much of the conflict that racked the miners' unions. Crises could pull the disparate individuals

---

133; Ross B. Moudy, "The Story of a Cripple Creek Miner," *Independent* 57 (August 18, 1904), 381; *Official Proceedings of the Ninth Annual Convention of the Western Federation of Miners of America, Denver, Colorado, May 27–June 6, 1901* (Pueblo, Colorado: Pueblo Courier Print, 1901), pp. 11–15; *Official Proceedings of the Tenth Annual Convention of the Western Federation of Miners of America, Denver, Colorado, May 26–June 7, 1902* (Denver: Colorado Chronicle, 1902), pp. 75–77; *Official Proceedings of the Eleventh Annual Convention of the Western Federation of Miners of America, Denver, Colorado, May 25–June 10, 1903* (Denver: Western Newspaper Union, 1903), pp. 197–198, 214. The issue of insurance was raised intermittently thereafter, but no affirmative action was taken until the convention of 1928 (*Report of the Proceedings of the Twenty-fifth Convention of the International Union of Mine, Mill and Smelter Workers, Salt Lake City, Utah, August 6–13, 1928* [Salt Lake City: Allied Printing, 1928], p. 3).

[3] Advertisement for Dillon Miners' Union #189, WFM, *Dillon* (Wyoming) *Doublejack*, May 13, 1905. The reports of W. E. Giese's Leadville operatives and C. L. Knuckey, alias P. B., detail miners' discussions of local, state, and national politics, the Boer War, the America's Cup races, character traits of Leadville's varied nationalities, and local celebrations (John F. Campion Papers, Western History Collection, University of Colorado Libraries, Boulder [hereafter cited as CU]). Union officials explored the practicality of a women's auxiliary during the December meeting of the Executive Board (Minutes of the executive board, Western Federation of Miners, December 1, 2, and 3, 1902, International Union of Mine, Mill and Smelter Workers Papers, CU).

and factions together, but such alliances were invariably temporary. Ethnic difficulties and jealousies proved recurrent saboteurs of unity. Melvyn Dubofsky's assertion that "while in some mining districts the foreign-born outnumbered the native Americans, no great ethnic division separated foreigners from natives" requires serious modification. Ethnic differences and national rivalries existed and became more serious as time passed. For example, while Dubofsky has assumed that Cornish-Irish enmity had disappeared by the end of the century, evidence from Leadville suggests that it continued to hamper miners' cooperation. The Cornish, who were largely unorganized, referred to the Cloud City Miners' Union as "the Irish outfit," while the Irish maintained that the Cornish were lackeys of the owners.

American and Irish miners were suspicious of other immigrants, and Eastern Europeans feared both the older immigrants and the Mexican-Americans. While the WFM professed class solidarity, its local leaders were leery of ethnic and racial minorities. Union organizers struggled with little success to eliminate or at least ameliorate these prejudices. Hard-rock miners also had a lingering suspicion of coal and lead miners, who were often used as strikebreakers. The fragmentary evidence suggests that hostilities of this sort continued to hamper union growth and development throughout the period.[4]

Unions were also deliberately weakened by mineowners' use of undercover agents. Since unions advocated universal membership, it was easy for detectives and dishonest individuals to secure responsible positions in which they could cripple the union or foster antagonism. Leadville's union was torn asunder by charges of espionage during the unsuccessful strike of 1896–1897. One faction alleged that John Ahern, a former union officer, had worked as a company spotter; his friends defended him, charging his accusers with obfuscation to cover their own wrongdoing. As late as 1902, the local union, the annual convention, and the executive board were all still attempting to sort through the recriminations. Though the executive board eventually

[4] Melvyn Dubofsky, "The Origins of Western Working Class Radicalism, 1890–1905," *Labor History* 7 (Spring, 1966), 135; reports of W. E. Giese's Leadville Operative, October 22 and October 24–25, 1899, Campion Papers, CU; *Proceedings of WFM, 1903*, p. 96; "Arizona Miners Join A.F.L. Union," *Miners' Magazine* 18 (November, 1917), 6; E. J. Dewar to J. B. McGinnis, Leadville, October 8, 1896, Gold Hill Miners' Union Papers, Bancroft Library of University of California, Berkeley (hereafter cited as BL).

found Ahern guilty and reprimanded him, the matter had festered and frustrated efforts to rebuild union strength in Leadville, and Campion and other managers had maintained their successful infiltration of the union by other agents.[5]

Embezzlement proved equally serious; local treasuries were stripped and unions failed when dishonest members fled with union assets. While some of the thieves were just ordinary miners unable to resist temptation, an occasional enterprising soul like Secretary Nylund of Telluride not only stole funds but also spied for the mine operators. While Nylund profited from this double deceit, the Telluride local reeled in confusion and then languished. Even Charles Moyer, long-time president of the Federation, was charged by the organization's 1928 convention with misappropriation of union funds and official malfeasance. WFM convention resolutions enjoined locals to bond treasurers, while the Virginia City Miners' Union obtained reliable and conscientious treasurers and secretaries by reserving these posts for disabled miners.[6]

Sometimes union men found themselves threatened by organized miners in neighboring camps or even in their own community. When weak unions in remote company towns were unable to resist management's pressures for wage reductions or longer hours, their capitulation jeopardized stronger unions in neighboring communities. In 1901 the Silver City, Idaho, union reacted quickly to allegations that the union miners in Como, Nevada, worked for wages below the union standard. Writing to their colleagues in Gold Hill, they suggested that their organizations join with Virginia City to investigate the Como situation.[7]

Most of the time, however, union miners cooperated with their colleagues in other locals and other unions by responding to appeals

[5] The details of the Ahern incident are discussed in reports of P. B., April 27, June 1, 8, 15, 22, and 29, and July 13, 1900, Campion Papers, CU; *Proceedings of WFM, 1902*, pp. 122–124, 125–126; minutes of the executive board, WFM, November 28, 1902, Mine, Mill Union Papers, CU.

[6] C. R. Waters to Charles E. Mahoney, Telluride, October 14, 1909, Mine, Mill Union Papers, CU; *Proceedings of Mine, Mill Union, 1928*, pp. 2–3; *Proceedings of WFM, 1901*, pp. 28, 58; *Proceedings of WFM, 1902*, p. 129; John Taylor Waldorf, *A Kid on the Comstock: Reminiscences of a Virginia City Childhood*, ed. Dolores Bryant Waldorf, p. 11.

[7] Armstrong to Officers and Members of Gold Hill Miners' Union, Silver City, August 18, 1901, Gold Hill Union Papers, BL.

for assistance and heeding requests not to come to communities with labor troubles, actual or anticipated. Routine assistance included helping newcomers find jobs. Miners joined in boycotting firms that were hostile to unions, as when the WFM convention of 1903 adopted a resolution to boycott the Stetson Hat Company. Furthermore, mining locals encouraged people to buy products with union labels and often joined state federations of labor in cooperative campaigns to promote unions among nonminers. They established a tradition of cooperation that extended from the individual and his local to the national federation.[8]

Miners were ordinarily practical men who organized because they believed that united they might better cope with the uncertainty that plagued their lives. As Frank Crampton explained: "The I.W.W. of that time, and the Western Federation of Miners did a fine job for the working stiffs. . . . Mine owners were hard, calloused, almost ruthless, and organized. . . . There were no safety laws, workmen's compensation, or state and federal legislation that protected the working stiffs." Wages, hours, and working conditions propelled miners into unions and remained central issues in the union movement, as an extant copy of the Gold Hill Miners' Union annual report of 1913 clearly indicates.[9]

Throughout the late nineteenth century miners sought to preserve the wage scales established during each camp's boom period. In an era of prolonged deflation, a constant wage meant an actual increase in purchasing power. When profits declined, as they did in the silver industry during the 1890's, management forced wage reductions or simply shut down. In the twentieth century, the situation changed; the economy entered an inflationary cycle, and the miners then needed wage increases to keep pace with a rising cost of living. But the owners seized traditional union demands for wage stability and turned them against the miners.

Closely linked to the subject of wages was the matter of hours.

[8] Bernard Coyle to Gold Hill Miners' Union, Eureka, Nevada, April 6, 1889, Gold Hill Union Papers, BL; reports of P. B., February 9, March 30, June 4, and July 13, 1900, Campion Papers, CU; *Proceedings of WFM, 1903*, pp. 152, 205.

[9] Frank A. Crampton, *Deep Enough: A Working Stiff in the Western Mine Camps*, pp. 11–12; WFM Annual Report of the Gold Hill Miners' Union, 1913, Gold Hill Union Papers, BL.

During the 1890's miners pressured owners and legislatures to limit shifts to eight hours. The eight-hour shift had existed in some mines since the 1860's, but other miners often worked ten or twelve hours underground. As unions tried to reduce the workday, managers claimed to see a conspiracy in which miners sought an effective wage increase. When not simply opposing eight-hour laws or union demands for an eight-hour shift, owners tried to maintain the prior wage-per-hour rate, whatever the number of hours worked. Conflicts over hours continued into the twentieth century and became an important issue in strikes and lockouts throughout the American West.

In addition to wages and hours, the issue of formal union recognition became a matter of grave concern. Miners credited whatever limited victories they won against the captains of the mining industry to their organization. Miners believed that according unions status as bargaining agents would give legitimacy to contracts and to their unions. Furthermore, management's formal recognition of unions would legitimize the active intervention of national or regional union representatives like Ed Boyce, Charles Moyer, and William Haywood. Management categorically opposed such recognition, claiming that unions were nothing more than a forum for outside agitators and that any recognition granted a union was an abdication of the legitimate rights of ownership.[10]

Unions championed several other causes as well, including the miners' desires to exercise some control over their working conditions. Locals occasionally demanded the rigid enforcement of local, state, or federal mine-safety legislation. They wanted an established procedure for handling grievances and asked for a voice in supervising company hospitals and medical benefits. Everywhere, union men sought assurances they would not be blacklisted. As early as 1864 Virginia City's union miners pressed for a closed shop as the only effective means of eliminating managerial reprisals for union involvement and membership. To the mid-twentieth-century mind most of

---

[10] Report of Gold Hill Miners' Union Committee Charged with Assessing the Propriety of Affiliation with the Coeur d' Alene Miners' Unions, January 8, 1891, Gold Hill Union Papers, BL; D. P. McGinley to Editor, *Miners' Magazine* 1 (April, 1900), 31; Leo J. Keena, "Cripple Creek in 1900," p. 3, Leo J. Keena Papers, State Historical Society of Colorado, Denver (hereafter cited as SHSC).

their requests seem reasonable, or at least "negotiable," but owners then believed the demands a threat to their investments, their control of the mines, and the American way of life.[11]

In this precarious environment, the WFM provided leadership and stability. It scrupulously followed its constitutional commitments to promote associations where none existed and to appoint and encourage organizers who would aid struggling or striking locals. The WFM executive board tried to discourage ill-conceived strikes and to inaugurate greater communication among individual unions. The board's communication efforts included attempts to develop a uniform code for transmitting and receiving sensitive information. The WFM encouraged the various locals to maintain listings of spies, scabs, detectives, delinquent members, nonaffiliated miners, and all corporations, owners, and managers who opposed unionization. Secretary William Haywood even proposed the creation of a union-at-large to serve miners in remote, unorganized regions. Finally, the Federation retained professional legal counsel to defend members charged with crimes, to fight injunctions, and to oppose the imposition of martial law in labor disputes. Though individuals occasionally grumbled about the "damnable assessments" that funded these activities, most members seemed to appreciate the central direction the national office provided.[12]

As an instrument of policy, the strike was always a two-edged sword in the miners' hands. They correctly perceived that halting all mining operations could imperil owners' profits and, exhausting the owners' patience, culminate in a union victory. Fear of profit losses did explain why operators in Virginia City, Butte, and Cripple Creek were willing to make important concessions to their workers. But the

---

[11] Walker De Marquis Wyman, Jr., "The Underground Miner, 1860–1910: Labor and Industrial Change in the Northern Rockies," (Ph.D. dissertation, University of Washington, 1971), pp. 133, 152; Joseph D. Cannon, "The Grievances at Miami and Globe, Arizona," *Miners' Magazine* 18 (August, 1917), 1; "Industrial Situation Tense in Globe and Miami Field," *Miners' Magazine* 18 (July, 1917), 1; Lingenfelter, *Hardrock Miners*, pp. 39–41.

[12] *Proceedings of WFM, 1901*, pp. 54, 57, 68; report of W. E. Giese's Leadville Operative, August 13, 1898, report of P. B., April 13, 1900, Campion Papers, CU; Philip Taft, "The Bisbee Deportation," *Labor History* 13 (Winter, 1972), 7–8; *Proceedings of WFM, 1902*, pp. 75, 111; *Proceedings of WFM, 1903*, pp. 38, 151–152, 167–168.

union overestimated the immediate impact of an actual strike on owners and underestimated the attendant hardships for union members. Few miners' unions consciously built up their financial resources in preparation for a strike. Individually, the men generally lacked the accumulated savings that would carry them and their families through the prolonged idleness. In contrast, management rarely faced actual privation, regardless of how long a strike lasted.

The Leadville miners' strike of 1896–1897 provides an instructive example of the problems miners encountered when they walked out. On June 19, 1896, the Cloud City Miners' Union struck all mines that refused to restore the three-dollar-per-day wage. By early July, some workmen began to have second thoughts about continuing the strike. While praising his union, one miner complained that he had worked only eight days prior to the strike and had expected to apply for his "traveling card" so that he could leave Leadville in good standing and find work elsewhere. Edward Grant, miner, reportedly felt that the decision to strike had been "too rash and hasty." Men without savings suffered, and all tired of the dull routine of manning the picket lines. Intent on breaking the union, managers finally used troops and strikebreakers to undermine the laborers' unity. Writing of his involvement in the Goldfield strike of 1907, J. C. Kinnear described the nagging foreboding that individuals experienced on the eve of a work stoppage. "Being a member of the Western Federation of Miners, I had to quit like all others. I well remember going home the last afternoon, facing a cold wind and with light snow blowing in my face. Not a very bright prospect without a job and very little money."[13]

While general concerns like wages, hours, and working condi-

[13] Reports of Pinkerton Operative A. C. H., Leadville, July 2, 3, and 6 and August 7 and 9, 1896, Eben Smith to John F. Campion, Denver, September 22, 1896, Campion Papers, CU; John L. McNeil to Max Boehmer, Leadville, September 8, 1896, Wolftone Consolidated Mining and Milling Company Papers, SHSC; J. J. Blow to Mac, Leadville, October 4, 1896, J. J. Blow Papers, CU; J. C. Kinnear, "Fifty Years in Mining, 1907–1957," p. 16, J. C. Kinnear Papers, Western History Research Center, University of Wyoming, Laramie (hereafter cited as WHRC). Whereas the "rustling cards" mentioned in chapter 7 were issued by the companies to miners in or coming to the community, the "traveling cards" mentioned here were issued by the local unions to members in good standing who were leaving the district and thus looking for work elsewhere.

tions were the critical labor issues, men occasionally struck from visceral indignation at a particular management policy. Such action might be nothing more than the firing of drunken employees, which S. W. Mudd alleged had precipitated a strike at Leadville's Baldwin Mine. Especially in remote camps where miners' dependence on the company seemed dramatically emphasized, miners objected to corporate policies that forced bachelors to reside in company-owned bunkhouses and to purchase board agreements at the company mess hall. The miners objected to what they perceived as exploitation and corporate paternalism in a free society.[14]

As noted in the preceding chapter, management's attempts to curtail high-grading precipitated labor conflict in Cripple Creek and Goldfield. Leo J. Keena remembered a Cripple Creek strike in 1900 over the Venture corporation's attempt to introduce the "Kimberly system," in which miners stripped naked and paraded past company supervisors. After a two-week struggle, the English managers restored "the old and time honored inspection basis of shaking and feeling the weight of the dinner buckets of the out-going shift." In Goldfield the introduction of the change house triggered labor discord that lingered from 1906 to 1908.[15]

There is also circumstantial evidence to suggest that managers occasionally "manufactured" strikes, as in the previously noted Chrysolite affair. It is certain that corporations took advantage of strikes, even to the point of stockpiling minerals for sale in case of labor trouble. In June, 1896, Karl Eilers asked J. J. Blow, his Leadville associate, about the prospects for a strike. "I am anxious to know about that strike of the miners. Does there seem to be any prospect of it at present? I am getting a very great amount of sulphides daily and if there is to be no strike I shall want to cut this off somewhat." In 1912 the *Salt Lake City Herald-Republican* interviewed "a prominent mining engineer" who claimed that the Bingham Canyon strike had come "at a most opportune time" since "extensive electrification of the two big concentrators, the steam shovels and the railroad itself

---

[14] Mudd to John F. Campion, Leadville, September 20, 1899, Campion Papers, CU; Vernon H. Jensen, *Heritage of Conflict: Labor Relations in the Non-Ferrous Metal Industry up to 1930*, pp. 140–141.

[15] Keena, "Cripple Creek," p. 4, Keena Papers, SHSC; Mrs. Hugh Brown, *Lady in Boomtown: Miners and Manners on the Nevada Frontier*, p. 68.

is contemplated for the immediate or near future. This can be accomplished now at minimum cost and without the inconveniences of attempting operations on the usual extensive scale while inaugurating the changes from steam power to electricity."[16]

Managers, with some justification, blamed labor agitators for provoking strikes. At various times, managers' notions of agitators included all union members, WFM organizers, walking delegates of union locals, or any Federation official who happened into town during the course of a strike. As members of an industrial union preaching against wage and hour disparities as well as local grievances, the labor activists and the national representatives did become agents of change. In that sense, they were agitators, arousing the discontent that often produced strikes. However, when the managers characterized these union demands as alien, radical, and impractical, they were simply dismissing the legitimate complaints of industrial workers.[17]

Strikes could sometimes be averted if there was a mutual commitment to negotiation and compromise. D. C. Copley, WFM executive board member, explained to the convention of 1903 how he had mediated a dispute between miners in Bonanza, Colorado, and the Hanover Mining and Milling Company. By encouraging the parties to elaborate their respective positions carefully, he was able to gain mutual concessions that restored industrial harmony. Miners' decisions to strike generally followed a pattern; miners met, drafted their grievances, and then carried them to the mine where they presented their complaints and called the workers out. Rarely willing to negotiate without consulting their superiors, the company officials either treated the workers' delegation respectfully but noncommittally or refused all contact with them. When a strike was prolonged, the owners

[16] Karl Eilers to J. J. Blow, June 11, 1896, Blow Papers, CU; "Doubts Ability of Moyer to Best Utah Copper's Manager," *Salt Lake City Herald-Republican*, September 21, 1912, Strike Scrapbook 1, Daniel Cowan Jackling Papers, BL.

[17] John F. Campion to E. C. Simmons, Denver, January 28, 1898, Campion Papers, CU; Marshall Sprague, *Money Mountain: The Story of Cripple Creek Gold*, pp. 249–258; "Moyer Coming to Direct Miners," and "Agitators Gettin' in Their Work at Tintic Gist of Message," *Salt Lake City Herald-Republican*, September 12 and 20, 1912, Strike Scrapbook 1, Jackling Papers, BL; Thomas A. Rickard, "Goldfield, Nevada—IV: Rich Ore and Its Moral Effects," *Mining and Scientific Press* 96 (1908), 776–777.

typically refused to bargain with the union, claiming instead a willingness to deal with the men individually.[18]

Miners, understanding the importance of good community relations when facing a protracted strike, often entered the contest with a reservoir of community good will. As John Calderwood demonstrated during the Cripple Creek strike of 1894, miners had to remain open to compromise, to maintain the support of sympathizers, and above all to establish an orderly and responsible public posture. Often one of the strikers' first acts was to forswear alcoholic beverages in order to curtail drunkenness and rowdyism and to conserve limited fiscal resources. Usually, members pledged themselves to remain peaceable and law-abiding, thereby defining for the community the parameters of the strike. Nonetheless, in prolonged struggles, it proved difficult to retain community sympathy, both because miners lacked money with which to buy goods from their friends and because management blamed the local depression on the strike and the union.[19]

As a strike progressed, the pressure on individual miners increased and, if the strike lasted too long, led to cracks in unity. When desperate enough, men reluctantly became scabs to feed themselves and their families. As J. C. Kinnear remembered it, "I don't like to record the fact I went to work as a strikebreaker, but it is the truth. . . . I only made this decision after it got to the point that Bertha and I didn't have enough to eat and we had rent to pay on the house every

[18] *Proceedings of WFM, 1903*, p. 115; Wm. F. McDowell to John F. Campion, Denver, 1897, Campion Papers, CU. Traditional emphasis on the incidence of violence by WFM unions has been tentatively challenged by James C. Foster, "Quantification and the Western Federation," *Historical Methods Newsletter* 10 (Fall, 1977), 141–148. Foster (p. 143) reports that "the mean number of violent, labor-related outbreaks was only 0.785 incidents in the lifetime [his estimate is about seven years] of the average local [he counts almost 600 locals in the United States and Canada]." Until Foster establishes a correlation between the size of the local and incidence of violence, and notes the percentage of the total membership affiliated with the largest locals, his findings must remain suggestive rather than definitive.

[19] Herbert G. Gutman, "The Workers' Search for Power: Labor in the Gilded Age," in H. Wayne Morgan, ed., *The Gilded Age: A Reappraisal*, pp. 38–68; W. Turrentine Jackson, *Treasure Hill: Portrait of a Silver Mining Camp*, pp. 130–136; Sprague, *Money Mountain*, pp. 133–156; reports of Pinkerton Operative A. C. H., Leadville, July 1, 4, and 6, 1896, Campion Papers, CU.

month." The union men maligned local and imported strikebreakers in song and verse. They scribbled off trite poetic invectives that found a ready forum in the *Miners' Magazine* under such scintillating titles as "What Scabs Won't Do."[20]

Management's use of professional detectives to infiltrate unions and to protect reopened mines and their work forces has already been discussed. Robert Livermore described these private guards as "nearly identical with the so-called 'bad men' of western legend, except that they were on the side of law and order." Owners also sought and often obtained support from state and national governments. Governors and the president had the power to declare martial law in communities where local officials proved unable to maintain peace and guarantee security, especially security of property. As early as the Leadville strike of 1880, the declaration of martial law proved a decisive factor in the owners' efforts to break the strike. In 1896 Leadville operator J. J. Blow described martial law as "the only way to settle this strike as the entire set of civil authorities, Judges, Sheriff, and most of the City police, all are in sympathy with the strikers." Martial law usually benefited the owners, as Blow anticipated, but when Colorado's governor, Davis H. Waite, used state militia during the 1894 Cripple Creek strike, it had the opposite effect, bringing the miners an important labor victory. Waite gained the confidence of the workers by impartially disarming both the strikers and the corporate deputies and then proceeded to mediate the controversy, asserting that "I believe this can be settled, if there is a disposition on the part of the mine owners to be conciliatory in these matters."[21]

More limited than mineowners in the tactics available to them, miners turned to the device of the sympathy strike. Before the creation of the WFM, individual unions could only appeal for moral and financial support from other miners' associations. After the formation of the Federation, though, the sympathy strike of WFM locals ap-

[20] Interview with Albert J. Frantz, Denver, April 6, 1972; Kinnear, "Mining," pp. 16–17, Kinnear Papers, WHRC; "What Scabs Won't Do," *Miners' Magazine* 18 (February, 1917), 6.

[21] Robert Livermore, "An Autobiography," pp. 75, 82, 126–127, Robert Livermore Papers, WHRC; Duane A. Smith, *Horace Tabor: His Life and the Legend*, pp. 136–139; J. J. Blow to C. G. Vicary, Leadville, December 10, 1896, Blow Papers, CU; "Transcript of Meeting of Mine Owners Committee with Governor Waite," June 2, 1894, John James Hagerman Papers, WHRC.

peared to offer a new way of pressuring either individual corporations or entire mining districts; it seemed particularly attractive in strikes against national conglomerates that had interstate holdings. The sympathy strike, however, proved a risky undertaking, since it could imperil strong unions in behalf of weak ones. The Federation's attorney, John H. Murphy, maintained that "the sympathetic strike does not accomplish very great results; that it is not as potential a factor in winning labor victories as the average man seems to think, and from my point of view I could see only disaster for the W.F.M. without any benefit."[22]

Frustrated strikers sometimes became desperate enough to consider the merits of rifle clubs and violence. They intimidated and harassed unorganized men, scabs, and intransigent managers. For example, Eliot Lord described how a Virginia City miners' delegation seized foreman John Trembath and attached him to a hoisting cable, upon which they raised and lowered him head first until he acceded to their demands. In Ely, Nevada, a similar assemblage advised operator John Traylor that they regarded him as an "undesirable resident" and intended to evict him. Traylor rejected their verdict and fired upon the group, killing three men and wounding several others. Such incidents, coupled with the spectacular allegations of Harry Orchard during the Haywood-Moyer-Pettibone trial, stamped the unions as violent in the public's eyes.[23]

Violence proved as inadequate a remedy for miners' ills as had the sympathy strike, so miners in search of an effective alternative turned to political action. By the early twentieth century, miners like T. H. Eckles had begun to blame their problems not on individual owners, but on the "infernal industrial system universally practiced by that 'person without a soul,' the modern mining corporation." At

[22] "Strike to Extend to Ely Mine," *Wichita Beacon*, September 20, 1912, "Moyer Will Call Sympathy Strike," *Salt Lake City Tribune*, September 20, 1912, both in Strike Scrapbook 1, Jackling Papers, BL; *Proceedings of WFM, 1903*, p. 161.

[23] Reports of P. B., October 23, 1899, May 8 and 11 and June 1 and 29, 1900, Campion Papers, CU; Moudy, "Cripple Creek Miner," pp. 381–382; Eliot Lord, *Comstock Mining and Miners*, pp. 183–184; Mark L. Requa, "The Nevada Consolidated Period," pp. 22–25, Mark Requa Papers, WHRC; Albert E. Horsley [Harry Orchard], *The Confessions and Autobiography of Harry Orchard*.

the local level, miners ran for sheriff or constable, while the Federation itself gradually switched its endorsement from the traditional parties to the Socialists. Evidence from Leadville suggests that the union leaders encouraged political activity and supported men who had distinguished themselves as friends of labor, men like D. C. Coates, Democratic candidate for lieutenant governor, and Frank W. Owers, Fifth District Judge. Frank Owers's electoral victories, which spanned over a decade, showed the effectiveness of union support at the local level. Elected first in the Populist sweep of 1894, Owers had opposed the imposition of martial law during the labor strike of 1896, had quashed injunctions against miners, and had routinely assisted in the naturalization of foreign miners. Leadville's operators regarded him as a "baneful and pernicious influence" and adopted assorted schemes to unseat him. They even went so far as to initiate a quo warranto proceeding against him, alleging that he did not reside within the Fifth Judicial District at all times.[24]

One Cripple Creek miner believed that the political consciousness of the unions in that district was one of their good points. He noted that bills affecting miners were regularly discussed in union meetings and that men were encouraged to oppose all unsatisfactory or discriminatory legislation. In Leadville the union's political activities included support of the state eight-hour-day referendum of 1905 and attempts to have a union member appointed deputy mine inspector for that area. In 1896 the union had pledged $200 to the campaign of William Jennings Bryan. On the Comstock, the miners supported state legislation regulating hours and working conditions in the mines. That union also backed an abortive attempt directed by Jacob Waldorf, miner and former president of the Virginia City Miners' Union, to legislate railroad rates. In 1907 the Gold Hill Miners' Union actively campaigned against federal legislation that would have cost miners jobs by eliminating or postponing the assessment work required for owners to maintain their claims.[25]

[24] T. H. Eckles, "Randsburg Comment," *Miners' Magazine* 1 (October, 1900), 35; "Three Grand Victories," *Miners' Magazine* 1 (December, 1900), 6; reports of P. B., September 21, October 24, and November 9, 1900, and June 4, 1901, J. C. Buck to S. W. Mudd, Denver, November 1, 1901, J. H. Weddle to John F. Campion, September 26, 1901, Kenneth L. Fahnestock to Senator Dexter T. Sapp, Leadville, March 30, 1905, Campion Papers, CU.

[25] Moudy, "Cripple Creek Miner," p. 382; reports of P. B., October 5,

This enthusiasm for politics flowed naturally from the locals through their delegates to the national conventions of the WFM. Federation officers believed that united action on important political issues could effect change, and in 1901 the WFM enacted resolutions opposing territorial expansion, federal intervention in labor disputes, the use of injunctions, an enlarged army, and the continued presence of Asiatics within the continental United States. The Federation also endorsed income, inheritance, and real land value taxes, postal savings banks, land laws that reserved the public lands for actual settlement, and the political reforms of initiative, referendum, and direct primaries. Two years later the WFM condemned war as "a scheme to enthrone privilege, to make the rich richer, the poor poorer, and in which the property of the rich receives more consideration than the lives of the poor." In 1918 the earlier stand against war was reversed, but the union now opposed so-called vagrancy laws, which required men to hold full-time employment, join the armed forces, or face imprisonment. These vagrancy laws, which drew wide support throughout the country, were frequently invoked against union organizers.[26]

The Western Federation's most controversial political stand, which came in the early years of the new century, began with a drift to the left and a renunciation of traditional parties. In 1901 the annual convention had endorsed political action that would be independent of the Republican, Democratic, or Populist parties, which it found "in the control of the capitalists." In 1902 President Ed Boyce moved a step further left, calling on the delegates to define a policy that would place the worker in opposition to "corporate oppression." On June 4, 1902, Leo Kenney made the historic motion that changed both the political position of the WFM and its image within the American labor movement. Adopted by a roll call vote of 129 1/6 yea, 70 1/3 nay, and 5 1/5 abstain, it read: "We, the tenth annual

November 9 and 16, 1900, and June 5, 1901, Campion Papers, CU; Waldorf, *Kid on Comstock*, pp. 11–18; Gold Hill Miners' Union No. 54 to Western Federation of Miners, November 25, 1908; Gold Hill Union Papers, BL.

[26] *Proceedings of WFM, 1901*, pp. 90–92; *Proceedings of WFM, 1902*, p. 135; *Proceedings of WFM, 1903*, pp. 13, 207; *Official Proceedings of the Twenty-third Convention and Third Biennial Convention of the International Union of Mine, Mill and Smelter Workers, 1918*, pp. 11–13; "Work-or-Fight Law Vetoed," *Miners' Magazine* 19 (August, 1918), 8.

convention of the Western Federation of Miners, do declare for a policy of independent political action, and do advise and recommend the adoption of the Socialist Party of America by the locals of the Federation in conjunction with a vigorous policy of education along the lines of political economy." From the beginning, the decision prompted internal discord, both ideological and personal, as name-calling began. Opponents of Socialist affiliation like B. P. Smith resented aspersions cast on them by pro-Socialists like Big Bill Haywood. The split between the trade unionist bloc and the supporters of socialism persisted from that time, and the annual meetings became the ideological battleground between the factions.[27]

When political activism evolved into a national commitment to an alternative economic and political system, the Western Federation of Miners injected discord into its locals. Like sympathy strikes and violence, this stance branded the WFM as a radical organization. For the local leaders and the working members, these issues increased the difficulties of obtaining concerted cooperation among a transient and ethnically splintered mining population. That they were to any degree able to overcome these handicaps and coordinate efforts for the improvement of local conditions and the amelioration of grievances is a tribute to their intelligence, industry, and ability to recognize problems as shared.

[27] *Proceedings of WFM, 1901*, pp. 89, 106; *Proceedings of WFM, 1902*, pp. 9, 74–75, 94–96, 99–100, 104.

9.

"Only a Man in Overalls,
Lay Him Anywhere—"

MINING opportunities lured men into the boom towns of the Rocky Mountains and Great Basin. For six decades they came, and, once there, they followed the latest news of bonanza and borrasca. They called themselves hard-rock miners, and they gouged out nature's mineral wealth. In the process, they adjusted to a hostile environment, erected camps and towns, experienced joy and sorrow, learned the tasks of mining, and endured the industrial wage earner's plight. One marvels at the adaptability and ingenuity of these people. Consciously or unconsciously, these men and their families were American pioneers; by dint of their efforts, they created new societies on which they engrafted their civilization.

Ray Allen Billington has provided a reference point for comparing miners with other groups of western pioneers. In *America's Frontier Heritage*, Billington describes the traits and attitudes that he believes developed from the American frontier experience. Specifically, he focuses on the mobility—spatial, social, and economic—typical of frontiersmen. Looking at the intermountain miners from this same perspective highlights their similarity to other western pioneers.

The geographic mobility of the mining population, described earlier, was perhaps the most typical feature of miners' lives. Miners and their families came from around the world and sought work wherever the mines were operating. Population shifts from established mining towns to remote outposts accompanied each successive strike of new ore. Miners flocked from Virginia City to the White Pine District, Austin, and Pioche in the late 1860's. Subsequent generations left Central City and Georgetown bound for Leadville in 1879–1880, Leadville for Cripple Creek and Creede in the early 1890's, and Cripple Creek for Goldfield, Tonopah, and the copper camps of Arizona, Nevada, and Utah in the early twentieth century. Coming west either as "seekers" or "escapees" (Billington uses the terms "abundancy" and "deficiency" motivated), miners severed traditional ties and accepted the uncertainty of western mining. In search of personal satisfaction, they adapted to the necessity of movement, whether they liked it or not. Some did like it, and, like the snow-bird miners, displayed a real wanderlust, seeking out frequent changes in scenery and working conditions. Others, who simply sought an elusive job security and a good wage, joined the current not out of love for change but in pursuit of personal progress and achievement.

Geographic mobility also characterized other western laborers. Farm workers, lumberjacks, cowboys, sheepherders, railroad men, soldiers, and even civil servants migrated or were moved from place to place in the nineteenth- and twentieth-century American West. Like miners, they too possessed the option of quitting one job to find something more satisfactory elsewhere. Miners, then, were typical of other Western workers in their freedom of movement and their willingness to engage in repeated migration.

Billington sees pioneers' geographic movement as linked to the desire for social and economic mobility. Duane A. Smith makes a similar point in his article, "Colorado's Urban-Mining Safety Valve." He argues that the expansion of Colorado's mining frontier prior to the 1890's not only provided employment for all willing to work but also sustained hopes for improvement and created options for the discontented. According to Smith, the general absence of labor discord in Colorado's hard-rock mining camps before 1890 supports his view of miners as optimistic about advancement and suggests that the mining frontier acted as a "safety valve" in Colorado. During the 1890's, the decline in new discoveries, the maturing of existing camps, the repeal of the Sherman Silver Purchase Act, and the Panic of 1893 combined to limit the opportunities available to wage-earning miners.[1]

How real and how extensive miners' opportunities for social and economic mobility ever were on the frontier is a complex issue, however. Many historians have asserted that opportunities for upward movement abounded, but only recently have handwritten census returns, local tax and probate records, and church, fraternal, and association records been used to shed light on the reality. Billington has

[1] Ray Allen Billington, *America's Frontier Heritage*, pp. vii, 23–46, 181–198; Duane A. Smith, "Colorado's Urban-Mining Safety Valve," *Colorado Magazine* 48 (Fall, 1971), 299–318.

suggested that, in any event, not the fact of mobility but the belief in it was crucial in conditioning the frontier experience. Believing that equality and opportunity existed in their society, the early frontier residents acted on that belief and thereby created them to a degree. Perceptions have their own reality and may reshape the human response to situations.[2]

The evidence indicates that the miners considered themselves full and equal members of their communities and stood shoulder to shoulder with their fellow citizens in times of crisis. For example, miners joined in defending their towns from the ravages of fire and disease, and they took no pleasure from the fact that capitalists had to close their mines. They stood for the equality of all white men and abhorred pretentiousness. Their famed snipe hunts, badger fights, and shivarees served both as initiation ceremonies and as expressions of the equality of men. The belief that every man was as good as his neighbor also helps explain why condescending eastern preachers had a rugged time amidst the miners, who wanted to be treated as individuals and as free men. For the immigrants who made up nearly 50 percent of this work force, mere treatment as equals represented real social mobility compared with what they had known. The frustration of aspirations to full equality, though, helps explain the outrage shown by union members who found themselves treated arbitrarily by an employer or manager.

Many miners, though, did find a better life in the mines than they had known elsewhere, or than they would likely have found. There is undeniable evidence that some miners actually did become prominent owners and managers. For example, Bingham Canyon's Louis Buchman, a Russian immigrant, rose from the lowly position of mucker earning a mere $2.50 a day to that of manager. Similarly, Tom Kearns of nearby Park City was transformed by a lucky strike from miner and occasional prospector to one of Utah's leading business and political leaders. Those miners who did achieve success served as examples and symbols of what men could achieve and thus fueled the belief in the mining frontier as a land of opportunity.[3]

[2] Billington, *Frontier Heritage*, pp. 97–116, 139–180.
[3] Russell G. Frazier, "Bingham Canyon through the Eyes of a Company Doctor," *Utah Historical Quarterly* 33 (Fall, 1965), 284; Vernon H. Jensen, *Heritage of Conflict: Labor Relations in the Non-Ferrous Metal Industry up to*

Sometimes his start in mining opened to a man opportunities in other, related occupations. For example, William J. Penrose started mining as an eleven-year-old boy in Cornwall. After a career that took him to mines in Pennsylvania and Illinois, Penrose moved west and mined in Tuscarora and Ruby Hill, Nevada. While in Ruby Hill, he worked as a reporter on the local paper and eventually became editor and proprietor of the *Ruby Hill Mining News.* After he moved to Butte, Montana, in 1885, he established another newspaper, the *Butte Mining Journal.* With this base of support he entered politics and was elected to the Montana House of Representatives in 1889. Ironically, his career ended with an assassin's bullet after he opposed an eight-hour law that his constituents supported. Virginia City's Father Patrick Manogue was a former miner whose California mining experiences aided his ministry to the Roman Catholics on the Comstock. He established a strong parish among his co-religionists and probably contributed to Eliot Lord's conclusion that "it is worth noting that the clergy who complain least of the apathy of the miners are the Catholic priests, who take pains to come nearer to their parishioners than the Protestant ministers are able or willing to do."[4]

Other men who had served a stint in the mines attained higher economic and social status in later life. Both Bernard Baruch and Jack Dempsey mucked as youths in the mines of Cripple Creek. Ross B. Moudy worked as both miner and smelterman before he garnered a position as assistant state chemist of Wyoming. The attainments Moudy listed for Cripple Creek union leaders were typical accomplishments for enterprising miners. Characterizing the union heads as young and energetic, Moudy noted that they had become successful leasers prior to the Cripple Creek strike of 1903–1904.[5]

1930, p. 98; Richard Hermann Peterson, "The Bonanza Kings: Mining Entrepreneurs of the Trans-Mississippi Frontier" (Ph.D. dissertation, University of California, Davis, 1971), pp. 40–42; Richard E. Lingenfelter, *The Hardrock Miners: A History of the Mining Labor Movement in the American West, 1863–1893,* pp. 189–192.

[4] John Bernard McGloin, S.J., "Patrick Manogue, Gold Miner and Bishop," *Nevada Historical Society Quarterly* 14 (Summer, 1971), 27–28; Eliot Lord, *Comstock Mining and Miners,* p. 406.

[5] Ross B. Moudy, "The Story of a Cripple Creek Miner," *Independent* 57 (August 18, 1904), 380–382.

Individuals like J. C. Kinnear, Frank Crampton, and George W. Parsons combined ambition with education, social background, or good connections to escape the lot of a lifelong miner. Kinnear was rescued from a Goldfield winze by a former employer, who promptly sent him off to Diamondfield to supervise the Daisy Mine. Crampton, the son of a middle-class family, had received a good education before his dismissal from an ivy-league college. His education and his personal perseverance in the correspondence study of assaying and surveying provided him with the background necessary for his career as a surveyor, assayer, prospector, and mining engineer. George Parsons came from a middle-class, white-collar background and soon linked himself to the professional and financial elite of early Tombstone. These men represented a category of miners who were intelligent, capable, and, more importantly, well-born, a condition that facilitated their transitions from miners to managers or entrepreneurs.[6]

The pursuit of individual goals provided some men with their personal versions of economic and social success. Tuscarora's Edward M. Lafferty worked intermittently as a wage miner, a bartender, a gambler, and a shopkeeper's clerk in order to continue prospecting on his own properties. Always in need of grubstakes, he worked at whatever job was available. Cripple Creek's Johnson Barbee struggled along on the money earned from his small claims in the hope that someday he would make a big strike and thereby raise the money needed for his daughter's education and his own retirement. Cornish miners saved their meager wages and attained lasting fame as prosperous, frugal men. Widely respected for both their skill and industry, the Cousin Jacks frequently obtained promotions to shift bosses, stationary engineers, or positions of minor supervisory responsibility.[7]

[6] J. C. Kinnear, "Fifty Years in Mining, 1907–1957," J. C. Kinnear Papers, Western History Research Center, University of Wyoming, Laramie; Frank Crampton, *Deep Enough: A Working Stiff in the Western Mine Camps*, pp. 59, 87, 152–155; George Whitwell Parsons, *The Private Journal of George Whitwell Parsons*, July 8 and September 4, 1880, pp. 136, 159; Peterson, "Bonanza Kings," p. 45.

[7] "Statement of Edward Mathew Lafferty, 1881," p. 10, Edward Mathew Lafferty Papers, Bancroft Library of University of California, Berkeley; Mabel Barbee Lee, *Cripple Creek Days*, pp. 166–167; Lynn I. Perrigo, "The Cornish

As these examples show, there was some occupational mobility for miners. Not all of it, though, was upward. Erstwhile miners took many jobs, some of which were described in the first chapter. The miners did not always succeed at their new occupations, and some of them returned to the mines. Of these, some had simply been failures; others had had trouble adjusting to the humdrum existence of storekeepers, farmers, liverymen, and public officials. Though most miners had neither the opportunity nor, perhaps, the inclination for social and economic advancement, witnessing the accomplishments of others reinforced their belief in individualism, equality, and opportunity.

Miners' experiences and perceptions of mobility fit fairly well Billington's characterization of other American frontiersmen. But the frontier experience was only one aspect of the forces that shaped the miners' lives and environment. Mining life was also urban and resembled urban life elsewhere in the United States. Furthermore, the dependence of the mining industry on external sources of investment tied the mining West closely into the national economy.

Broad macroeconomic trends affected the mining frontier just as they influenced developments in New York, Chicago, New Orleans, and San Francisco. Business cycles brought periods of relative prosperity interspersed with depressions, falling wages, and unemployment. The Panic of 1893 was especially severe in the silver camps of the mining West. The prolonged deflation of the late nineteenth century increased the purchasing power of miners' wages, just as it did those of other wage earners. Miners did not, perhaps, suffer from modernization as much as other workers, because mining had always been a somewhat industrial occupation. Miners were spared some of the ongoing transitions, further divisions of labor, and extensive mechanization that characterized industries like shoemaking. In fact, improved technology was slow in altering the actual mechanics of mining, but it quickly spurred productivity. Dynamite, steam hoists, air drills, and electric haulage systems did for the miners' productivity what stronger rails, better engines, and larger cars did for the railroad

Miners of Early Gilpin County," *Colorado Magazine* 14 (May, 1937), 94; Arthur Cecil Todd, *The Cornish Miner in America*, p. 162; D. Smith, "Safety Valve," pp. 311, 312, 316; W. Turrentine Jackson, *Treasure Hill: Portrait of a Silver Mining Camp*, pp. 227–228.

worker, or what Bessemer and open-hearth processes did for steel-workers.

When individual mineowners or corporations consolidated their holdings and bought out rival companies, they were responding to waste, inefficiency, and the profit motive in a way that the Rocke-fellers, Morgans, and Carnegies employed and applauded. Increasing-ly, corporations replaced individual capitalists, and this too paralleled national trends. As a result, though, the miners discovered that ab-sentee ownership complicated their presentations of grievances. Work-ing miners became increasingly alienated and class conscious, and, like other wage earners, they distrusted the corporations and their representatives. As a result, they forged local unions to meet local problems and created the Western Federation of Miners to coordi-nate their various local efforts. The tactics of labor confrontation in the mining industry were identical to those utilized by carpenters, plumbers, textile workers, shoemakers, coal miners, and steelworkers. Wages, hours, and union representation were the most common causes of labor discord. Angry corporate executives denounced labor associations and used ironclad contracts, blacklisting, lockouts, strike-breakers, and detectives to counter union activity, especially strikes or threatened strikes. Union members responded in kind, often at-tempting to intimidate strikebreakers, detectives, and employers. At the national level, the Citizens' Alliances that union men so despised were aided and promoted by the National Association of Manufac-turers, the Citizens' Industrial Association, the American Anti-Boycott Association, and the National Council for Industrial Defense. These organizations assisted in the so-called employer counteroffensive in the face of early twentieth-century union gains.[8]

In most respects, miners' living conditions compared favorably with those of other industrial and agricultural wage earners. Their urban problems paralleled those of other city residents: fires, diseases,

[8] W. Elliot Brownlee, *Dynamics of Ascent: A History of the American Economy*, p. 191; Edward C. Kirkland, *Industry Comes of Age: Business, Labor and Public Policy, 1860–1897*, pp. 325–409; John A. Garraty, *The New Commonwealth, 1877–1890*, pp. 78–178; Harold U. Faulkner, *The Decline of Laissez Faire, 1897–1917*, pp. 249–314; Glenn Porter, *The Rise of Big Busi-ness, 1860–1910*; Melvyn Dubofsky, *Industrialism and the American Worker, 1865–1920*.

crime, and inadequate sanitation, housing, and public services. Not only did these problems plague other townsmen of the period, but they have done so throughout history. In their relatively smaller cities and towns, the miners often had a better chance to effect meaningful changes in the quality of life than their counterparts who lived and worked in the major metropolitan centers. The miners' wages compared favorably with those of skilled tradesmen and were far superior to those paid unskilled factory hands and agricultural laborers. While contracting and leasing occasionally incorporated features of piece-work, the speedup, and the stretch-out, all of which plagued the textile industry, miners could and did refuse to accept contracts that they felt would not "make wages." The accommodations available to the miners may have left something to be desired, but they were superior to the crowded, vermin-infested, unsanitary tenements that awaited many workers in the major cities. Throughout most of the period, 1860–1920, hard-rock mining was a semiskilled occupation whose western practitioners were optimistic and proud. Whether they were right or not, they believed they could shape their own destinies, and they struggled to rear their families and to provide their children with amenities they had lived without. Miners confronted head-on the forces that shaped contemporary America: the frontier, capitalism, industrialization, and urbanization. Perhaps a fitting eulogy for them was simply that they led the forces of their sometimes destructive and wayward civilization into the American hinterland.

Present-Day Mining
by P. J. Holohan

Well, Fred, you asked me for to tell "what we do in the mine";
According to your ideas our lot is rather fine,
Because we do not work long hours from daylight until dark,
You have a notion in your head our life is all a lark.

In ranching, as you call it, the hours are long, you say:
In springtime ploughing and discing, in summer shoveling hay.
But allow me to inform you, the back is always safe;
There is no chance for sloughing ground, to put you in your grave.

I will try to tell you, if I can, how we spend the hours
That we're immured in living tombs, from sunshine and from showers.
Whilst you enjoy the blessings of the sun's refreshing beams
And Nature's gifts so lavish strewn in valleys, dales and streams.

There's the man who runs the wiggle-tail [rock drill] and swallows all
    the dust;
They never think about him when they prattle "safety fust";
But as my forte's not medicine, I cannot really tell
How dust affects a miner's lungs—perhaps it's just as well.

No doubt it helps the boarding boss, for when a man is full
Of concentrates of lead and zinc his appetite is dull—
Especially since the state went dry, and one can get no booze
To use it as a pinch-bar for to jar the blamed thing loose.

Yet the man who runs the buzzy "in the shadow of the bomb"
Is generally gay and hearty and a stranger unto gloom.
Don't talk of heroes in a trench, who fight the masters' war,
The miners with their ratchets are a braver lot by far.

There's the mucker with his No. 2, perhaps a No. 4—
He shovels rock into the box from off a swayback floor.
Should the boulders be too massive, he can rest himself quite well
By pounding with a ten-pound sledge, for they're as hard as h——l!

NOTE: First published in *The Miners' Magazine* 17 (April 6, 1916), 5.

There's the trammer with his jitney, which is sometimes off the track;
He lifts until he strains himself in trying to put it back.
At last he finds he has not strength to lift a ton of ore,
And then starts out to look for help; he should have done before.

There's the timberman and helper; they surely need some strength
To wrestle logs some two feet through and nine times that in length,
Which have been laying in some creek since Noah quit the ark,
And weigh some eighty pounds a foot—that is, without the bark.

They hoist them by a dinky and they land them on a floor,
They drag them through the grassy stopes until the bones are sore;
They "horse" them over muck piles with the most apparent ease;
The men are surely religious; they work upon their knees.

And when they get them to the place, each man has got a stitch,
But they're placed from foot to hanging without a single hitch.
There's the cage and the skinner and the man who runs the pump;
The pipeman with his stilson-sure, he's always on the jump.

He carries with him tools enough to stock a hardware store—
From monkey wrench to chain tongs and of valves a half a score.
And should you need a piece of pipe, no matter what's the size,
He'll put it in so quickly it will take you by surprise.

There's the nipper as he packs the steel—sometimes he has a pick;
He's as busy as a nailer as he does the double quick.
The load he totes would kill a mule, but mules they cost some pluck,
While nippers, when they get too slow, can be thrown out as junk.

There's the shaftman in the bottom, with a drill called a bull moose;
A piston drill is not too slow to break the ground up loose;
The day of two men on a drill is gone and passed away;
One man does two men's work right now, but does not draw the pay.

Now you know about the working bunch—the men who toil and sweat;
The reason why the hours are eight, I have not told you yet.
If we were still unorganized as ranch hands are today,
The hours would sure be ten or twelve, with very little pay.

The boon of shorter hours was won by economic power
Displayed by men in unity who never yet did cower,
And though they did not always win, they never wholly lost;
When we compare the gains they made, they minimize the cost.

If those who work upon a ranch would simply organize
And stand as one together, they would be acting wise.
And when the union's formed be active for its good;
Do not start into knocking and go searching for "new blood."[1]

[1] This closing reference to "new blood" alluded to a controversy within the Western Federation of Miners. The minority of members attracted to the IWW and its philosophy sought to oust the conservative leadership of Charles Moyer and W. L. Burns. Though ultimately unsuccessful, the "new blood" movement rocked the WFM between 1916 and 1918.

Accidents and Injuries Typical in Hard-Rock Mining in the Intermountain West Prior to 1920

TABLE 1. Percentages of Accidents Attributable
to Specific Causes on the Comstock (1863–1880)
and in the Arizona Copper Company Mines
(1913–1919)

Cause of Accidents	Comstock (1863–1880)*	Arizona Copper Company (1913–1919)**
Falling objects	16.2%	49.7%
Caving, roof falls	10.3	0.5
Explosives	7.7	0.3
Falls	22.8	4.6
Machinery (not hoisting or haulage)	3.2	16.1
Cages	24.6	0.4
Cars	—	13.2
Gas/heat prostration	2.0	0.6
Burns/hot water	0.6	0.3
Fires	6.2	—
Miscellaneous	6.4	14.4

SOURCE: Eliot Lord, *Comstock Mining and Miners,* p. 414; six-month reports of serious accidents, 1913–1919, Arizona Copper Company Ltd. Papers, Special Collections Department, University of Arizona Library, Tucson. Figures for Arizona Copper Company do not total 100% due to rounding.
  *n 901
  **n 3,887

TABLE 2. Common Injuries as Percentage of Total Accidents
in Arizona Copper Company Mines, 1914–1919

Part of Body	1914	1915	1916*	1917	1918	1919
Arm, hand, finger	39.4	32.3	31.5	34.5	35.5	34.3
Leg, foot, toe	25.5	20.6	22.2	16.9	18.1	21.8
Head	13.0	19.9	14.8	16.5	15.4	12.5
Back	6.5	4.7	6.7	6.3	10.2	6.5
Eye	3.2	5.0	4.0	9.4	11.7	14.2

SOURCE: Six-month reports of serious accidents, 1914–1919, Arizona Copper
Company Ltd. Papers, Special Collections Department, University of Arizona
Library, Tucson.
　　*Statistics for 1916 covered only the six-month period April 1–September
30, 1916. Data on the preceding six months were not available in the Arizona
Copper Company Collection.

TABLE 3. Accident Rate per 1,000 Men Employed
by Arizona Copper Company, 1913–1919

Accidents (Per Thousand)	1913	1914	1915	1916	1917*	1918	1919
Killed	12.5	7.9	7.1	2.7	4.7	5.4	4.0
Seriously Injured	266.8**	136.8	116.9	82.8	123.4	77.6	166.0
Slightly Injured		288.8	271.4	284.0	343.1	394.5	398.1
TOTAL ACCIDENTS	279.3	433.5	395.4	369.5	471.2	477.5	568.1

SOURCE: Six-month reports of serious accidents, 1913–1919, Arizona Copper
Company Ltd. Papers, Special Collections Department, University of Arizona
Library, Tucson.
　　*In 1917 the mines of the Arizona Copper Company were shut down for
three months. The accident rate per 1,000 men employed has been adjusted to
reflect this work stoppage.
　　**In 1913 figures for the slightly injured and the seriously injured were
not reported separately. This figure is their combined total.

TABLE 4. Number, Rate, and Causes of Metal-Mine
Accidents in the Intermountain West, 1911–1919

	1911	1912	1913	1914	1915	1916	1917	1918	1919
Number of Accidents	4,175	6,269	6,761	7,217	7,098	10,425	10,693	12,521	7,449
Rate per Thousand Employed	150	203	172	264	308	282	304	410	324
Underground									
Fall of rock ore	46.9%	38.7%	33.7%	31.1%	30.5%	27.4%	28.6%	25.9%	26.3%
Fall of rock/ore while loading at face	—	3.6	5.7	8.2	6.9	9.6	9.2	11.1	10.6
Timber and hand tools	10.3	12.8	14.9	12.3	10.0	9.8	9.4	8.4	9.1
Explosives	2.5	3.5	2.0	2.6	1.6	2.7	1.9	1.7	2.3
Haulage	10.9	11.6	13.8	12.5	12.8	13.2	13.0	13.1	12.5
Falling down chute, raise, etc.	5.9	5.4	5.5	3.8	3.9	3.6	4.0	3.9	3.0
Run of ore from chute/pocket	1.5	2.0	3.6	1.9	2.1	4.3	1.0	3.3	2.5
Drilling	4.7	4.8	4.2	6.4	6.9	5.0	5.0	6.2	5.5
Electricity	0.5	0.4	0.3	0.3	0.3	0.4	0.3	0.4	0.3
Machinery (drills & locomotives excepted)	2.3	3.0	1.0	0.7	0.5	0.5	0.7	1.0	0.9
Fires	0.9	0.1	0.1	0.1	—	0.1	—	—	0.1
Suffocation from natural gases	0.2	0.3	0.2	0.1	0.2	0.2	0.2	0.2	0.4
Inrush of water*	—	—	—	—	—	—	—	—	—
Stepping on nails	—	1.9	1.6	1.2	2.4	2.2	2.4	2.8	2.4
Other miscellaneous	9.1	9.9	11.2	16.5	19.4	18.4	21.6	20.0	20.6
TOTAL	95.5	97.9	97.9	97.7	97.5	97.3	97.4	98.0	96.4

TABLE 4. *continued*

	1911	1912	1913	1914	1915	1916	1917	1918	1919
Shaft									
Falling down shaft	0.9	0.5	0.4	0.3	0.4	0.4	0.3	0.2	0.5
Objects falling down shaft	1.3	0.5	0.4	0.4	0.5	0.7	0.4	0.4	0.7
Breaking of cables	—	0.1	—	0.1	—	—	0.1	0.1	0.2
Overwinding	0.3	—	0.1	—	—	—	—	—	—
Skip or cage related	—	0.7	0.8	0.8	0.9	0.8	0.9	0.8	0.8
Other miscellaneous	1.6	0.3	0.4	0.6	0.7	0.8	0.9	0.5	1.3
TOTAL	4.1	2.1	2.1	2.3	2.5	2.7	2.6	2.0	3.6

SOURCE: Albert H. Fay, comp., "Metal-Mine Accidents in the United States . . ." for calendar years 1911–1918, and William W. Adams, "Metal-Mine Accidents in the United States during the Calendar Year 1919," U.S. Bureau of Mines, *Technical Papers* 40, 61, 94, 129, 168, 202, 224, 252, and 286 (1913–1921). Figures are for Arizona, Colorado, Nevada, New Mexico, Utah, and Wyoming. The Bureau of Mines reported fatalities broken down by the specific metals mined, and only the statistics for copper and gold and miscellaneous metals (including silver) were included in the compilations for this table. In the 1911 report only, the same categories were used in reporting injuries and the same procedure was followed in compiling this table. Figures for subsequent years include injuries in all metal mines in the region.

*There were a few such accidents in each of the years reported but less than 0.1 percent of all accidents. Throughout the period (and even today) the threat from underground water continued to require special precautions in mining.

Bibliography

Archival Collections

Albuquerque, New Mexico. Zimmerman Library of University of New Mexico. Special Collections Department.
Blumenthal Papers. Holm O. Bursum Papers.
Berkeley, California. Bancroft Library of University of California.
Ancient Order of United Workmen, Virginia City Lodge No. 1 Papers. James M. Copeland Papers. Gold Hill Miners' Union Papers. Edward Robbins Howe Papers. Daniel Cowan Jackling Papers. Edward Mathew Lafferty Papers. Tom T. Lane Papers. Henry H. Mason Papers. Thomas Waters Papers.
Boulder, Colorado. University of Colorado Libraries. Western History Collection.
J. J. Blow Papers. John F. Campion Papers. Gregory Consolidated Mines Papers. George H. Holt Papers. International Union of Mine, Mill and Smelter Workers (formerly the Western Federation of Miners of America) Papers. Mary McKinney Mining Company Papers. Hal Sayre Papers. Arthur Wifley Papers.
Denver, Colorado. Denver Public Library. Western History Department.
Diary of Arthur Bowen. Ernest LeNeve Foster Papers. George H. and Mary Galbraith Papers. Frank Hall Papers. William A. Hamill Papers. Diary of George O. James. Mary Murphy Gold Mining Company Papers. Diary of Cecil C. Morgan. George Rowe Papers. Eben Smith Papers. Ezra M. Stahl Papers. Stevens Mining Company Papers. Theodore F. Van Wagenen Papers. Wm [William] Letters.
Denver, Colorado. State Historical Society of Colorado.
Oren Allen Papers. Sylvester Allen Papers. Diary of Ella Baily. Diary of I. J. Baldwin. Alfred P. Camp Papers. Thomas F. Dawson Scrapbooks. James Vila Dexter Papers. Eagle River Consolidated Mining Company Papers. Albert S. Frost Papers. Lizzie Gaynor Papers. Edwin Harrison Papers. Nathaniel P. Hill Papers. Marmaduke B. Holt Papers. Leo J. Keena Papers. Robert Orchard Old Papers. John Marshall Paul Papers. Jesse Summers Randall Papers. Records of Colorado Mining Districts. Albert Byron Sanford Papers. W. Frank

Witherell Papers. Wolftone Consolidated Mining and Milling Company Papers.

Laramie, Wyoming. University of Wyoming. Western History Research Center.

O. S. Alers Papers. Frank Ayer Papers. Daniel Moreau Barringer Papers. Marshall Bond Papers. John E. Burton Papers. Henry Collbran Papers. Alfred B. Colwell Papers. John B. Farish Papers. E. D. Gardner Papers. Goodrich Papers. H. V. S. Groesbeck Papers. James John Hagerman Papers. Grace Raymond Hebbard Papers. Ross Hoffman Papers. J. C. Kinnear Papers. Robert Livermore Papers. Mining File. Henry C. Morris Papers. Eben E. Olcott Papers. Edwin L. Regennitter Papers. Mark Requa Papers. Agnes Wright Spring Papers. Josiah Spurr Papers. Joel P. and John Parker Whitney Papers.

San Marino, California. Huntington Library.

Henry Douglas Bacon Papers. James Duncan Hague Papers. [Gideon] Anthony Hamilton Papers. Loring Family Papers. Nevada Mining Companies Papers. Isaac Lawrence Requa Papers. Sierra Nevada Silver Mining Company Papers. Utah Mining Papers. Henry Ellsworth Wood Papers.

Santa Fe, New Mexico. State Records Center of New Mexico.

L. Bradford Prince Papers. Rennehan-Gilbert Papers.

Tucson, Arizona. University of Arizona Library. Special Collections Department.

Arizona Copper Company, Ltd. Papers. Frances Douglas Papers. Diary of M. O'Gorman. Josephine Roche Papers. William Field Staunton Papers. Stoddard Family Papers. United Verde Copper Company Papers.

Interviews

Frantz, Albert J., Office of United Steelworkers of America, District 38, Denver, April 6, 1972, interviewed by author. Interviews of the Colorado Pioneers (PAM). Collected and prepared under the joint auspices of the State Historical Society of Colorado and the Works Projects Administration, summary transcripts held by the State Historical Society of Colorado, Denver.

Richards, Bessie Launder. "Mining Town Memories—Colorado and Mexico." Typescript of interview conducted by Mel Erskine, Berkeley, 1967. Bancroft Library, University of California, Berkeley.

Public Documents

Arizona Reports. 1866–1920.

Bureau of Labor Statistics of the State of Colorado. *Biennial Reports.* 1877–1920.

Colorado Reports. 1862–1920.

Commissioner of Labor of the State of Nevada. *Biennial Report.* 1915–1916.

Industrial Commission of Utah. *Report.* 1917–1920.

U.S. Bureau of Mines. *Bulletins.* 1910–1930.

———. *Technical Papers.* 1911–1930.

U.S. Congress, House. *Papers Relative to Labor Troubles at Goldfield, Nevada.* 60th Cong., 1st sess., 1908, House Doc. 607.

U.S. Congress, Senate. *A Report on Labor Disturbances in the State of Colorado from 1880 to 1904, Inclusive, with Correspondence Relating Thereto.* 58th Cong., 2nd sess., 1905, Senate Doc. 122.

———. *Review of the Labor Troubles in the Metalliferous Mines of the Rocky Mountain Region.* 58th Cong., 2nd sess., 1904, Senate Doc. 86.

———. *Statement of the Western Federation of Miners.* 58th Cong., 2nd sess., 1904, Senate Doc. 163.

U.S. Department of the Interior and Bureau of the Census. *Census Reports.* Eighth through Fourteenth. 1860–1920. Tables for occupations.

U.S. Industrial Commission. *Report.* 19 vols. 1900–1902.

Newspapers and Periodicals

Denver Rocky Mountain News, 1860–1861, 1890–1900.

Engineering and Mining Journal (New York), 1866–1932. Title is *American Journal of Mining, Milling, Oil Boring, Geology, Mineralogy, Metallurgy, etc.* until July, 1869.

Georgetown (Colorado) *Courier,* 1877–1878.

Miners' Magazine (Denver), 1900–1920.

Mining and Scientific Press (San Francisco), 1860–1922. Title is *Scientific Press* until 1872.

Books and Articles

Allen, James B. "The Changing Impact of Mining on the Economy of Twentieth Century Utah." *Utah Historical Quarterly* 38 (Summer, 1970), 240–255.

———. "The Company Town: A Passing Phase of Utah's Industrial Development." *Utah Historical Quarterly* 34 (Spring, 1966), 138–160.

———. *The Company Town in the American West.* Norman: University of Oklahoma Press, 1966.

"Among American Miners." *Chambers Journal* (ser. 5) 8 (September 5, 1891), 571–573.

Angel, Myron. *History of Nevada, with Illustrations and Biographical Sketches of Its Prominent Men and Pioneers.* 1881. Reprint. Berkeley: Howell-North, 1958.

Arrington, Leonard J. "Abundance from the Earth: The Beginnings of

Commercial Mining in Utah." *Utah Historical Quarterly* 31 (Summer, 1963), 192–219.

———. *The Changing Economic Structure of the Mountain West, 1850–1950*. Monograph Series, no. 10. Logan: Utah State University Press, 1963.

———. "Utah and the Depression of the 1890's." *Utah Historical Quarterly* 29 (January, 1961), 3–18.

Atherton, Lewis. "Structure and Balance in Western Mining History." *Huntington Library Quarterly* 30 (November, 1966), 55–84.

Baker, James H., and Leroy R. Hafen, eds. *History of Colorado*. 5 vols. Denver: Linderman Co., Inc., 1927.

Baker, Ray Stannard. "Organized Capital Challenges Organized Labor: The New Employers' Association Movement." *McClure's Magazine* 23 (July, 1904), 279–292.

———. "The Reign of Lawlessness: Anarchy and Despotism in Colorado." *McClure's Magazine* 23 (May, 1904), 43–57.

Bancroft, Hubert Howe. *The Works of Hubert Howe Bancroft*. 39 vols. San Francisco: The History Company, 1882–1890.

Bechtol, Paul T., Jr. "The 1880 Labor Dispute in Leadville." *Colorado Magazine* 47 (Autumn, 1970), 312–325.

Billington, Ray Allen. *America's Frontier Heritage*. Hinsdale, Illinois: The Dryden Press, 1966.

Bird, Leah M. "Minor Political Parties in Colorado." *Colorado Magazine* 19 (November, 1942), 208–213.

Brooks, Juanita. "Silver Reef." *Utah Historical Quarterly* 29 (July, 1961), 281–287.

Brophy, A. Blake. *Foundlings on the Frontier: Racial and Religious Conflict in Arizona Territory, 1904–1905*. Tucson: University of Arizona Press, 1972.

Brown, Mrs. Hugh. *Lady in Boomtown: Miners and Manners on the Nevada Frontier*. Palo Alto: American West Publishing Co., 1968.

Brown, Richard Maxwell. "The American Vigilante Tradition." In *Violence in America: Historical and Comparative Perspectives*, edited by Hugh Davis Graham and Ted Robert Gurr. New York: Bantam Books, 1969, pp. 154–226.

Brownlee, W. Elliot. *Dynamics of Ascent: A History of the American Economy*. New York: Alfred A. Knopf, 1974.

Buchanan, John W. "The History of a Ghost Town, Caribou." *Colorado Magazine* 21 (November, 1944), 201–207.

Bureau of the Census and Social Sciences Research Council. *The Statistical History of the United States from Colonial Times to the Present*. Stamford, Conn.: Fairfield Publishers, Inc., 1965.

Clum, John P. "It All Happened in Tombstone." *Arizona Historical Review* 2 (October, 1929), 46–72. Reprint edited by John D. Gilchriese. *Arizona and the West* 1 (Autumn, 1959), 232–247.

Cochran, Alice. "John Langrishe and the Theater of the Mining Frontier." *Colorado Magazine* 46 (Fall, 1969), 324–337.

Commons, John R., David J. Saposs, Helen L. Sumner, E. B. Mittelman, H. E. Hoagland, John B. Andrews, and Selig Perlman. *History of Labour in the United States.* 4 vols. New York: The Macmillan Company, 1926–1935.

Conlin, Joseph R. *Big Bill Haywood and the Radical Union Movement.* Syracuse: Syracuse University Press, 1969.

———. "The Haywood Case: An Enduring Riddle." *Pacific Northwest Quarterly* 59 (January, 1968), 23–32.

Conner, Daniel E. *Joseph Reddeford Walker and the Arizona Adventure.* Edited by Donald J. Berthrong and Odessa Davenport. Norman: University of Oklahoma Press, 1959.

Cornwall, Harry C. "My First Year in the Gunnison Country." Edited by Duane Vandenbusche. *Colorado Magazine* 46 (Summer, 1969), 220–244.

Crampton, Frank A. *Deep Enough: A Working Stiff in the Western Mine Camps.* Denver: Sage Books, 1956.

Currie, Barton Wood. "Housekeeping in Alkali Land." *Good Housekeeping*, August, 1909, pp. 119–128.

Davidson, Levette J. "The Festival of Mountain and Plain." *Colorado Magazine* 25 (July, 1948), 145–157; (September, 1948), 203–212.

Davies, J. Kenneth. "Utah Labor Before Statehood." *Utah Historical Quarterly* 34 (Summer, 1966), 202–217.

Davis, Carlyle C. *Olden Times in Colorado.* Los Angeles: The Phillips Publishing Co., 1916.

De Pew, Kathryn. "William A. Hamill." *Colorado Magazine* 32 (October, 1955), 266–279.

Derber, Milton. "The Idea of Industrial Democracy in America, 1898–1915." *Labor History* 7 (Fall, 1966), 259–286.

"The Diary of John McQuig." *Nevada Historical Society Quarterly* 6 (Spring, 1963), 3–27.

Digerness, Helen Sidney. "Jesse Summers Randall and Pioneer Georgetown." *Colorado Magazine* 22 (November, 1945), 258–265.

Drury, Wells. *An Editor on the Comstock Lode.* New York: Farrar & Rinehart, Inc., 1936.

Dubofsky, Melvyn. "The Leadville Strike of 1896–97: An Appraisal." *Mid-America* 48 (April, 1966), 99–118.

———. "The Origins of Western Working Class Radicalism, 1890–1905." *Labor History* 7 (Spring, 1966), 131–155.

———. *We Shall Be All: A History of the Industrial Workers of the World.* Chicago: Quadrangle Books, 1969.

Dwyer, Robert J. "The Irish in the Building of the Intermountain West." *Utah Historical Quarterly* 25 (July, 1957), 221–235.

Elliott, Russell R. *History of Nevada*. Lincoln: University of Nebraska Press, 1973.

————. *Nevada's Twentieth-Century Mining Boom: Tonopah, Goldfield, Ely*. Reno: University of Nevada Press, 1966.

Faulkner, Harold U. *The Decline of Laissez Faire, 1897–1917*. The Economic History of the United States, vol. 7. New York: Holt, Rinehart & Winston, Inc., 1951.

Foner, Philip S. *History of the Labor Movement in the United States*. 4 vols. New York: International Publishers Co., Inc., 1949–1965.

Fossett, Frank. *Colorado: Its Gold and Silver Mines, Farm and Stock Ranges, and Its Health and Pleasure Resorts*. New York: C. G. Crawford, 1880.

Foster, James C. "Quantification and the Western Federation." *Historical Methods Newsletter* 10 (Fall, 1977), 141–148.

Frazier, Russell G. "Bingham Canyon through the Eyes of a Company Doctor." *Utah Historical Quarterly* 33 (Fall, 1965), 283–288.

Friedman, Lawrence M. *A History of American Law*. New York: Simon and Schuster, 1973.

Fuller, Leon W. "Colorado's Revolt Against Capitalism." *Mississippi Valley Historical Review* 21 (December, 1934), 343–360.

————. "Governor Waite and His Silver Panacea." *Colorado Magazine* 10 (March, 1933), 41–47.

————. "A Populist Newspaper of the Nineties." *Colorado Magazine* 9 (May, 1932), 81–87.

Fulton, Carl. "The Winter of the Deep Snow." *Colorado Magazine* 40 (January, 1963), 38–41.

Galloway, John Debo. *Early Engineering Works Contributory to the Comstock*. University of Nevada *Bulletin* 41, no. 5. Geology and Mining Series.

Garraty, John A. *The New Commonwealth, 1877–1890*. New York: Harper & Row, Publishers, 1968.

Glaab, Charles M., and A. Theodore Brown. *A History of Urban America*. New York: The Macmillan Co., 1969.

Goodwin, J. M. "Snowslides in the Rockies: A Perilous Study." *Overland Monthly* (ser. 2) 29 (April, 1897), 379–386.

Green, Thomas L. "Junius R. Lewis and the Golden Chest Mining Company." *Colorado Magazine* 50 (Winter, 1973), 24–40.

Greever, William S. *The Bonanza West: The Story of the Western Mining Rushes, 1848–1900*. Norman: University of Oklahoma Press, 1963.

Gressley, Gene M., ed. *Bostonians and Bullion: The Journal of Robert Livermore, 1892–1915*. Lincoln: University of Nebraska Press, 1968.

Gressley, Gene M. *West by East: The American West in the Gilded Age*. Charles Redd Monographs in Western History, no. 1. Provo: Brigham Young University Press, 1972.

Griswold, Don L., and Jean H. Griswold. *The Carbonate Camp Called Leadville.* Denver: University of Denver Press, 1951.

Hall, Frank. *History of the State of Colorado.* 4 vols. Chicago: The Blakely Printing Co., 1889–1895.

Hand, William. "The Western Federation of Miners." *Outlook* 83 (May 19, 1906), 125–133.

Hanington, C. H. "Early Days of Central City." *Colorado Magazine* 19 (January, 1942), 3–14.

Harmon, E. M. "The Snow Slide at Toponis Mine, 1883." *Colorado Magazine* 17 (May, 1940), 115–117.

Harvey, Mrs. James R. "The Leadville Ice Palace of 1896." *Colorado Magazine* 17 (May, 1940), 49–101.

Hastings, Frank S. *A Ranchman's Recollections: An Autobiography.* Chicago: Sanders Publishing Co., 1921.

Hastings, James K. "A Boy's Eye View of the Old Southwest." *New Mexico Historical Review* 26 (October, 1951), 287–301.

Haywood, William D. *Bill Haywood's Book: The Autobiography of William D. Haywood.* New York: International Publishers, 1929.

Henderson, Charles W. *Mining in Colorado: A History of Discovery, Development and Production.* U.S. Geological Survey *Professional Paper* no. 138. Washington, D.C.: Government Printing Office, 1926.

Henderson, Patrick C. "The Bradshaw Bonanza." *New Mexico Historical Review* 38 (April, 1963), 151–162.

———. "The Peck Mine: Silver Bonanza in the Bradshaw Mountains." *Arizona and the West* 4 (Autumn, 1962), 227–236.

Holbrook, Stewart. *The Rocky Mountain Revolution.* New York: Henry Holt & Co., 1956.

Hollon, W. Eugene. *The Great American Desert: Then and Now.* New York: Oxford University Press, 1966.

Homsher, Lola M., ed. *South Pass, 1868: James Chisholm's Journal of the Wyoming Gold Rush, USA.* Lincoln: University of Nebraska Press, 1960.

Horsley, Albert E. [Harry Orchard]. *The Confessions and Autobiography of Harry Orchard.* New York: The McClure Co., 1907.

Hotchkiss, S. C. "Occupational Diseases in the Mining Industry." *American Labor Legislation Review* 2 (February, 1912), 131–139.

Hough, Merrill. "Leadville and the Western Federation of Miners." *Colorado Magazine* 49 (Winter, 1972), 19–34.

Ingham, G. Thomas. *Digging Gold among the Rockies.* Philadelphia: Cottage Library Publishing House, 1881.

Jackson, W. Turrentine. "British Impact on the Utah Mining Industry." *Utah Historical Quarterly* 31 (Fall, 1963), 347–375.

———. "The Infamous Emma Mine: A British Investment in the Little Cottonwood District, Utah Territory." *Utah Historical Quarterly* 23 (October, 1955), 339–362.

————. *Treasure Hill: Portrait of a Silver Mining Camp.* Tucson: University of Arizona Press, 1963.

Jensen, Vernon H. *Heritage of Conflict: Labor Relations in the Non-Ferrous Metal Industry up to 1930.* Ithaca: Cornell University Press, 1950.

Johnson, Rue C. "Theatre in Zion: The Bingham City Dramatic Association." *Utah Historical Quarterly* 33 (Summer, 1965), 187–197.

King, Alfred Castner. *Mountain Idylls and Other Poems.* Chicago: Fleming H. Revell Co., 1901.

King, Joseph E. *A Mine to Make a Mine: Financing the Colorado Mining Industry, 1859–1902.* College Station: Texas A&M University, 1977.

Kinikin, L. C. "Early Days in Telluride." *Colorado Magazine* 26 (January, 1949), 14–26.

Kirkland, Edward C. *Industry Comes of Age: Business, Labor and Public Policy, 1860–1897.* Chicago: Quadrangle Books, 1967.

Kluger, James R. *The Clifton-Morenci Strike: Labor Difficulties in Arizona, 1915–1916.* Tucson: University of Arizona Press, 1970.

Lamar, Howard Roberts. *The Far Southwest, 1846–1912: A Territorial History.* New York: W. W. Norton & Co., Inc., 1970.

Langdon, Emma F. *The Cripple Creek Strike, 1903–1904.* Victor, Colorado: Daily Record Press, 1904.

Lee, Mabel Barbee. *Cripple Creek Days.* New York: Doubleday & Co., Inc., 1958.

Leonard, Frank. "Helping the Unemployed in the Nineteenth Century: The Case of the American Tramp." *Social Service Review* 15 (December, 1966), 429–434.

Lewis, Oscar. *Silver Kings: The Lives and Times of MacKay, Fair, Flood, and O'Brien, Lords of the Comstock Lode.* New York: Alfred A. Knopf, 1947.

Lillard, Richard G., ed. "A Literate Woman in the Mines: The Diary of Rachael Haskell." *Mississippi Valley Historical Review* 31 (June, 1944), 81–98.

Lindquist, John H. "The Jerome Deportation of 1917." *Arizona and the West* 11 (Autumn, 1969), 233–246.

Lingenfelter, Richard E. *The Hardrock Miners: A History of the Mining Labor Movement in the American West, 1863–1893.* Berkeley: University of California Press, 1974.

Lonsdale, David L. "The Fight for an Eight-Hour Day." *Colorado Magazine* 43 (Fall, 1966), 339–353.

Lord, Eliot. *Comstock Mining and Miners.* 1883. Reprint. Berkeley: Howell-North, 1959.

Mayo, Guy. "Rawhide Pays Its Respects: Wherein Riley Grannan, King of the Sport of Kings, Runs His Last Furlong." *The American West* 10 (January, 1973), 40–47.

Mazzulla, Fred M. *Brass Checks and Red Lights*. Denver: privately printed, 1966.

McGloin, John Bernard, S.J. "Patrick Manogue, Gold Miner and Bishop." *Nevada Historical Society Quarterly* 14 (Summer, 1971), 25–31.

"Mine Explosions." *Scientific American Supplement*, March 29, 1902, pp. 21938–39.

"Mining Labor Union." *The Nation*, July 14, 1904, p. 31.

Moore, Richard T. "The Arizona Bureau of Mines." *Journal of the West* 10 (January, 1971), 136–141.

Morgan, Dale L., ed. *Rand McNally's Pioneer Atlas of the American West*. New York: Rand McNally & Co., 1969.

Morgan, H. Wayne, ed. *The Gilded Age: A Reappraisal*. Syracuse: Syracuse University Press, 1963.

Morris, Henry C. *Desert Gold and Total Prospecting*. Washington, D.C.: privately printed, 1955.

————. *The Mining West at the Turn of the Century*. Edited by C. S. Lewis. Privately printed, 1962.

Moudy, Ross B. "The Story of a Cripple Creek Miner." *Independent* 57 (August 18, 1904), 380–382.

Mumford, Lewis. *Technics and Civilization*. New York: Harcourt, Brace & World, 1963.

Murray, Robert A. "Miner's Delight, Investor's Despair: The Ups and Downs of a Sub-Marginal Mining Camp in Wyoming." *Annals of Wyoming* 44 (Spring, 1972), 25–56.

Nelson, Elroy. "The Mineral Industry: A Foundation of Utah's Economy." *Utah Historical Quarterly* 31 (Summer, 1963), 178–191.

"Nevada." *Harper's Weekly* 47 (June 20, 1903), 1017–34.

Neve, Mattie W. "The Winter of 'the Big Snow.'" *Colorado Magazine* 40 (January, 1963), 31–37.

Noxon, Victor I. "Hard Rock Drilling Contests in Colorado." Transcribed and prepared by Forest Crossen. *Colorado Magazine* 11 (May, 1934), 81–85.

Oakeshott, Gordon B. "Notes on the Nevada Bureau of Mines." *Journal of the West* 10 (January, 1971), 152–154.

Ourada, Patricia K. "The Chinese in Colorado." *Colorado Magazine* 29 (October, 1952), 273–284.

Pace, Josephine. "Kimberly as I Remember Her." *Utah Historical Quarterly* 35 (Spring, 1967), 112–120.

Palmer, Louise M. "How We Live in Nevada." *Overland Monthly* 2 (May, 1869), 457–462.

Papanikolas, Helen Zeese. "Life and Labor Among the Immigrants of Bingham Canyon." *Utah Historical Quarterly* 33 (Fall, 1965), 289–315.

————. "Toil and Rage in a New Land." *Utah Historical Quarterly* 38 (Spring, 1970), 100–203.

Parsons, George Whitwell. *The Private Journal of George Whitwell Parsons.* Phoenix: Arizona Statewide Archival and Records Project (W.P.A.), 1939.

Paul, Rodman W. *Mining Frontiers of the Far West, 1848–1880.* New York: Holt, Rinehart and Winston, Inc., 1963.

Pawar, Sheelwant B. "The Structure and Nature of Labor Unions in Utah, an Historical Perspective, 1890–1920." *Utah Historical Quarterly* 35 (Summer, 1967), 236–255.

Pendleton, Mark A. "Memories of Silver Reef." *Utah Historical Quarterly* 3 (October, 1930), 99–118.

"Peppler and McNamara's Miner's Safety Hat." *Scientific American*, May 2, 1896, p. 276.

Perrigo, Lynn I. "The Cornish Miners of Early Gilpin County." *Colorado Magazine* 14 (May, 1937), 92–101.

————. "The First Two Decades of Central City Theatricals." *Colorado Magazine* 11 (July, 1934), 141–152.

————. "Law and Order in Early Colorado Mining Camps." *Mississippi Valley Historical Review* 28 (June, 1941), 41–62.

Peterson, Richard H. *The Bonanza Kings: The Social Origins and Business Behavior of Western Mining Entrepreneurs, 1870–1900.* Lincoln: University of Nebraska Press, 1977.

————. "Conflict and Consensus: Labor Relations in Western Mining." *Journal of the West* 12 (January, 1973), 1–17.

Poet, S. E. "The Story of Tin Cup, Colorado." *Colorado Magazine* 9 (January, 1932), 30–38.

————. "Victor, Colorado—The City of Mines." *Colorado Magazine* 10 (May, 1933), 106–114.

Pomeroy, Earl. *The Pacific Slope: A History of California, Oregon, Washington, Idaho and Nevada.* New York: Alfred A. Knopf, 1965.

Porter, Glenn. *The Rise of Big Business, 1860–1910.* Arlington Heights, Illinois: AHM Publishing Corp., 1973.

Presser, E. Louise. "Memoirs of E. Louise Presser, Rhyolite, 1906–1908." *Nevada Historical Society Quarterly* 15 (Summer, 1972), 45–51.

"Prevention of Dust in Mines." *Scientific American*, April 28, 1900, p. 261.

Rastall, Benjamin M. *The Labor History of the Cripple Creek District: A Study in Industrial Evolution.* Madison: University of Wisconsin Press, 1908.

Rickard, Thomas A. "Gold Mining Activity in Colorado." *North American Review* 162 (April, 1896), 473–480.

————. "Goldfield, Nevada—IV: Rich Ore and Its Moral Effects." *Mining and Scientific Press* 96 (June 6, 1908), 774–777.

————. "Mining at High Altitude." *Cassier's Magazine* 22 (October, 1902), 688–698.

Robertson, Rhoda Fletcher. "The Snowslide at Tomichi." *Colorado Magazine* 12 (January, 1935), 35–37.

Rosen, George. *The History of Miners' Diseases: A Medical and Social Interpretation.* New York: Schuman's, 1943.

Rowse, A. L. *The Cousin Jacks: The Cornish in America.* New York: Charles Scribner's Sons, 1969.

Schoberlin, Melvin. *From Candles to Footlights: A Biography of the Pike's Peak Theatre, 1859–1876.* Denver: The Old West Publishing Co., 1941.

Shannon, David A. *The Socialist Party of America: A History.* New York: The Macmillan Co., 1955.

Shinn, Charles H. *Mining Camps: A Study in American Frontier Government.* New York: Charles Scribner's Sons, 1885.

———. *The Story of the Mine, as Illustrated by the Great Comstock Lode of Nevada.* New York: D. Appleton & Co., 1896.

Siringo, Charles A. *Two Evil Isms: Pinkertonism and Anarchism.* Austin: Steck-Vaughn Co., 1967. Facsimile reproduction of first edition, 1915.

Smith, Duane A. "The Caribou—A Forgotten Mine." *Colorado Magazine* 39 (January, 1962), 47–54.

———. "Colorado's Urban-Mining Safety Valve." *Colorado Magazine* 48 (Fall, 1971), 299–318.

———. *Horace Tabor: His Life and the Legend.* Boulder: Colorado Associated University Press, 1973.

———. "Mining Camps: Myth vs. Reality." *Colorado Magazine* 44 (Spring, 1967), 93–110.

———. *Rocky Mountain Mining Camps: The Urban Frontier.* Bloomington: Indiana University Press, 1967.

———. "The San Juaner: A Computerized Portrait." *Colorado Magazine* 52 (Spring, 1975), 137–152.

———. *Silver Saga: The Story of Caribou, Colorado.* Boulder: Pruett Publishing Company, 1974.

———. "The Vulture Mine: Arizona's Golden Mirage." *Arizona and the West* 14 (Autumn, 1972), 231–252.

Smith, Emory, Supervising Engineer. "Forty Years of Progress in Mine Safety." Reprinted from State of California, Department of Industrial Relations, Division of Industrial Safety, *Bulletin,* 1963.

Smith, Grant H. *The History of the Comstock Lode, 1850–1920.* University of Nevada *Bulletin* 37, no. 3. Geology and Mining Series.

"Snowshoe Post-Routes: From the *Colorado Graphic* of April 18, 1891." *Colorado Magazine* 17 (January, 1940), 36–38.

Snyder, Carl. "In the Cripple Creek Goldfields." *Harper's Weekly* 40 (1896), 57–59.

Spence, Clark C. *British Investments and the American Mining Frontier, 1860–1901.* Ithaca: Cornell University Press, 1958.

──────. "Knights of the Tie and Rail—Tramps and Hoboes in the West." *Western Historical Quarterly* 2 (January, 1971), 5–19.

──────. *Mining Engineers and the American West: The Lace-Boot Brigade, 1849–1933.* New Haven: Yale University Press, 1970.

Sprague, Marshall. *Money Mountain: The Story of Cripple Creek Gold.* Boston: Little, Brown & Co., 1953.

"State Workmen's Compensation Legislation." *American Labor Legislation Review* 2 (December, 1912), 565–567.

Stratton, David H. "The Cousins Jacks of Caribou." *Colorado Quarterly* 1 (Spring, 1953), 371–383.

──────. "The Rise and Decline of Caribou, Colorado." *Colorado Magazine* 30 (April, 1953), 109–118.

Street, Julian. "Colorado Springs and Cripple Creek." *Collier's*, November 21, 1914, p. 16.

Suggs, George G., Jr. "Catalyst for Industrial Change: The WFM, 1893–1903." *Colorado Magazine* 45 (Fall, 1968), 322–339.

──────. *Colorado's War on Militant Unionism: James H. Peabody and the Western Federation of Miners.* Detroit: Wayne State University Press, 1972.

──────. "Militant Western Labor Confronts the Hostile State: A Case Study." *Western Historical Quarterly* 2 (October, 1971), 385–400.

──────. "Prelude to Industrial Warfare: The Colorado City Strike." *Colorado Magazine* 44 (Summer, 1967), 241–262.

──────. "Strike-Breaking in Colorado: Governor James H. Peabody and the Telluride Strike, 1903–1904." *Journal of the West* 5 (October, 1966), 454–476.

Swanson, Evadene Burns. "Where's Manhattan?" *Colorado Magazine* 48 (Spring, 1971), 146–158.

Taft, Philip. "The Bisbee Deportation." *Labor History* 13 (Winter, 1972), 3–40.

Taylor, Bayard. *Colorado: A Summer Trip.* New York: G. P. Putnam & Sons, 1867.

Taylor, Robert Guilford. *Cripple Creek.* Indiana University Geographic Monograph Series, vol. 1. Bloomington: Indiana University Press, 1966.

Teetor, Col. H. D. "Mines and Miners around Idaho Springs." *Magazine of Western History* 12 (October, 1890), 667–671.

Teter, Wilburtine. "Sketches of Mining Life." *Harper's Weekly* 40 (1896), 59–62.

Thom, William B. "Early Day Marshalls of Tin Cup." *Colorado Magazine* 10 (July, 1933), 143–144.

Thompson, J. T. "Cousin Jack Stories." *Colorado Magazine* 35 (July, 1958), 187–192.

Todd, Arthur Cecil. *The Cornish Miner in America.* Glendale, Calif.: Arthur H. Clark Co., 1967.

Tondel, Frank P. "As I Remember Goldfield." *Nevada Historical Society Quarterly* 3 (Summer, 1960), 13–19.

Townley, John M. "The Delamar Boom: Development of a Small One-Company Mining District in the Great Basin." *Nevada Historical Society Quarterly* 15 (Spring, 1972), 2–19.

Trotter, E. J. *The Forbidden Fruit and the Prodigal Son.* Denver: privately printed, 1919.

Van Wagenen, Theodore F. "The Rights of the Miner." *Mining and Scientific Press* 96 (May 16, 1908), 669–673.

Waldorf, John Taylor. *A Kid on the Comstock: Reminiscences of a Virginia City Childhood.* Edited by Dolores Bryant Waldorf. Palo Alto: American West Publishing Co., 1970.

Walker, Arthur L. "Early Day Copper Mining in the Globe District." *Engineering and Mining Journal* 125 (April 14, 1928), 604–608; (April 28, 1928), 694–698.

———. "Recollections of Early Day Mining in Arizona." *Arizona Historical Review* 6 (April, 1935), 14–43.

Walling, William English. "The Labor 'Rebellion' in Colorado." *Independent* 57 (August 14, 1904), 376–379.

Walsh, Margaret. "The Census as an Accurate Source of Information: The Value of Mid-Nineteenth Century Manufacturing Returns." *Historical Methods Newsletter* 3 (September, 1970), 3–13.

———. "The Value of Mid-Nineteenth Century Manufacturing Returns: The Printed Census and Manuscript Census Compilations Compared." *Historical Methods Newsletter* 4 (March, 1971), 43–51.

Waters, Frank. *Midas of the Rockies.* Denver: Alan Swallow, Publisher, 1949.

Weinstein, James. "Big Business and the Origins of Workmen's Compensation." *Labor History* 8 (Spring, 1967), 156–174.

Wells, Merle W. "The Western Federation of Miners." *Journal of the West* 12 (January, 1973), 18–35.

West, Elliott. "Of Lager Beer and Sonorous Songs." *Colorado Magazine* 48 (Spring, 1972), 108–128.

Wiebe, Robert H. *The Search for Order, 1877–1920.* New York: Hill and Wang, 1967.

Williams, Joseph E., ed. *The Prentice-Hall World Atlas.* Englewood Cliffs, N.J.: Prentice-Hall, Inc., 1960.

Wolle, Muriel Sibell. "Ghosts in the Mountains." *Colorado Quarterly* 1 (Summer, 1952), 25–40.

———. "Irwin, a Ghost Town of the Elk Mountains." *Colorado Magazine* 24 (January, 1947), 8–15.

———. *Stampede to Timberline: The Ghost Towns and Mining Camps of Colorado.* Boulder: Alan Swallow, 1949.

Wright, William [Dan De Quille]. *The Big Bonanza.* 1876. Reprint. London: Eyre & Spottiswoode, 1969.

Wyman, Mark. "Industrial Revolution in the West: Hard-Rock Miners and the New Technology." *Western Historical Quarterly* 5 (January, 1974), 39–57.

Young, Otis E., Jr. *Western Mining: An Informal Account of the Precious-Metals Prospecting, Placering, Lode Mining, and Milling on the American Frontier from Spanish Times to 1893.* Norman: University of Oklahoma Press, 1970.

Unpublished Manuscripts

Byrkit, James Ward. "Life and Labor in Arizona, 1901–1921: With Particular Reference to the Deportations of 1917." Ph.D. dissertation, Claremont Graduate School, 1972.

Gallatin, E. L. "Reminiscences," typescript in State Historical Society of Colorado, Denver.

Hough, Charles M. "Leadville, Colorado, 1878–1898: A Study in Unionism." Master's thesis, University of Colorado, 1958.

King, Joseph Edward. "It Takes a Mine to Run a Mine: Financing Colorado's Precious-Metals Mining Industry, 1859–1902." Ph.D. dissertation, University of Illinois, 1971.

Lake, Henry F., Jr., ed. Remembrances of Pioneer Days in Gunnison County. Loose-leaf notebook of reprinted articles from the *Gunnison News-Champion*. Western History Research Center, University of Wyoming, Laramie.

Perrigo, Lynn I. "Life in Central City, Colorado, as Revealed by the *Register*: 1862–1872." Master's thesis, University of Colorado, 1934.

———. "Social History of Central City, Colorado, 1859–1900." Ph.D. dissertation, University of Colorado, 1936.

Peterson, Richard Hermann. "The Bonanza Kings: Mining Entrepreneurs of the Trans-Mississippi Frontier." Ph.D. dissertation, University of California, Davis, 1971.

Remington, John Alan. "Violence in Labor Disputes: The Haywood Trial." Master's thesis, University of Wyoming, 1965.

Rudolph, Gerald E. "The Chinese in Colorado, 1869–1911." Master's thesis, Denver University, 1964.

Sanders, Thomas D. and Charles C. Stemmer. "Hardships and Perils on the Western Frontier or Fifty Years of Pioneering." An autobiographical narrative dictated by Sanders to Stemmer, ca. 1924. Western History Department, Denver Public Library.

Smith, Duane Allan. "Silver Camp Called Caribou." Master's thesis, University of Colorado, 1961.

Suggs, George G., Jr. "Colorado Conservative versus Organized Labor: A Study of the James Hamilton Peabody Administration, 1903–1905." Ph.D. dissertation, University of Colorado, 1964.

Wyman, Walker De Marquis, Jr. (Mark). "The Underground Miner, 1860–1910: Labor and Industrial Change in the Northern Rockies." Ph.D. dissertation, University of Washington, 1971.

Index

Agassiz claim, 113
Ahern, John, 147–148
Alba City, Utah, 99
Alers, Major O. S., 14
Allen, James B., 24
Alley, Mrs. Margie, 77; husband of, 77
Alma Gold Mining and Milling Company, 113–114
American Anti-Boycott Association, 167
American Casualty Insurance and Security Company, 141
Americans, native-born: and hardrock mining, 8, 132–133; and highgrading, 120; and rivalries, 132, 147
America's Frontier Heritage, 161
Anaconda, Colo., 42
Ancient Order of United Workmen, 48
Anderson, A. J., 79, 111–112
Anderson, A. J., and Company, 111–112
Anderson, Charles, 122
Anderson, Grant, 85
animals, 41, 63; in mining, 65, 81, 82, 85, 95; in sports, 51, 53, 57
Animas River lode, 113
Antelope Mine, 98
Apex Mining Company, 109
Argall, Philip, 127–128
Arizona: climate in, 14; congress of, 40; mining statistics for, 11, 12, 13
Arizona Copper Company, Ltd., 77, 85
Arizona Legislature, 138
Arizona Supreme Court, 138
Arrington, Leonard, 88
Arthur, William, 114
Asiatics, 159. *See also* Chinese
Aspen, Colo., 32
asphyxiation, 87–88
assaying, 37, 116, 122; correspondence course in, 165; dishonesty in, 120–121
assumption of risk rule, 139
Austrians, 133, 134
avalanche, 17

bad air, 86–88
badger fight, 57–58
Balco, Fred, 77
Baldwin Mine, 153
Baltimore Mine, 103, 106
bankruptcy, 137
Barbee, Johnson (Jonce), 7, 97–98, 165
Barringer, Daniel M., 130, 134
Barringer, Lewin W., 119, 134
bars. *See* saloons
Baruch, Bernard, 164
baseball, 53
base metals, 21–22
Beckett, P. G., 105
bell signals, 83
Bennetts, John R., 113
Bibb, G. H., 113
Bickle brothers (miners), 90
Big Stick Mine, 41
Billington, Ray Allen, 161, 162–166
Bingham, Utah, 24
Bingham Canyon, Utah: shifts in, 63; strikes in, 133, 137, 153–154
Bisbee, Ariz., 26
Bison Mine, 98
Black Hawk, Colo., 102
blacklisting, 6, 122, 129, 150
blacks, 40
blacksmiths, 72
blasting, 70, 71–72, 86–87
blizzards, 15
Blow, J. J., 153, 156
B.O.B. Mining and Milling Co., 111–112
Bobtail Mine, 33, 66, 88
bonanza, 36

Bonanza, Colo., 154
borrasca, 36
Bouyer, Henry, 5
Bowers, Mrs. Sandy, 97
bowling, 53
boxing, 53
Boyce, Ed, 126, 150, 159
boycotts, 149
British Isles, 8–9, 132–133
Broad, _____ (miner), 76
Brown, Mrs. Hugh, 56, 72
Brown, Thomas, 88
Brownsville, Colo., 18, 34, 60, 102
Bryan, William Jennings, 159
Buchman, Louis, 163
Buckskin Joe, Colo., 10
Bulkley, Fred G., 104
burial, 91–92
Burns Detective Agency, 125
Burns, James T., 53, 123
burros, 63
Bursum, H. O., 112
Butler, Nev., 104
Butte, Mont., 89, 93, 151
Butte Mining Journal, 164

cage accidents, 82–84
Calderwood, John, 155
Camp, Alfred P., 7, 17
Campion, George, 134
Campion, John F.: detectives of, 132, 141; distrust of, for miners, 124, 142; endorses temperance, 39; ethnic prejudice of, 134; on high-grading, 120, 123; infiltrates union, 148; informers of, 124–125; names of mines of, 97–98; recommends workers, 143; on unions, 124
camps, 21–22
Canadian Society, 90
candles, 63; as accident cause, 85; fires caused by, 37, 89; for gas detection, 87; to light fuses, 72; as light in mines, 66–67; superstition about, 96
capitalism, 21–22, 130, 167, 168
Carbon County, Wyo., 133
carbon monoxide, 87–88
Carpenter, George A., 4
Carpenter, John, 113
Cash Entry Mine, 103
Catholics, 164
Catron, Thomas B., 113
cats, 41, 124
cave-ins, 77, 91, 92

celebrations, 49, 51
census. See United States Census Bureau; Colorado, census statistics
Central City, Colo.: childhood in, 32; effect of dust on men in, 93; ice skating in, 52; labor difficulties in, 10; opera in, 46; snowfall in, 15; wages in, 102
Central Pacific Railroad, 49
cerebrospinal meningitis, 42
Cerillos Mining District, N. Mex., 105
change houses, 122–123, 153
Cheyney, Sam. W., 130
children: on Independence Day, 51; games of, 52–53; gangs of, 39; loneliness of, 34; pastimes of, 33
Chinese, 9, 22, 30, 33, 159
cholera, 58
Christmas, 48, 50–51, 96
Chrysolite Mine, 131
circuses, 49
Citizens' Alliances, 167
Citizens' Industrial Association, 167
Civil War, 7
Clear Creek, Colo., 113–114
climate, 69
closed shop, 150
Cloud City Miners' Union, 48, 125, 152
coal miners, 147
coal mining, 23–24
Coates, D. C., 158
Cochise County, Ariz., 122, 143
Cochrane, Howard, 115
Coeur d'Alene, Idaho, 123
colds, 42
college students, 26
Colorado: census statistics, 9n, 11, 12, 13; climate, 14
Colorado Bureau of Labor Statistics, 101
Colorado City, Colo., 126
Colorado United Mining Company, 104
Colwell, Pearl O'Brien, 110
Colwell, Ray, 33, 48
Combination Mine, 118
Commerfield, Jack, 11, 93
communications, 83, 151
communities, urban nature of, 23, 36, 166, 167–168
commuting, 75
Como, Nev., 148
company towns, 23–26, 131–132
Comstock Lode, 65; accidents at, 77,

83; benefits of associations at, 145; change rooms at, 66; house at, 29; policies of officials of, 132; silver discovered in, 3; temperatures in mines at, 69; unemployment at, 127; wages at, 124
conservation, lack of, 16
Consolidated-Virginia Mine, 38
contests, 51–52, 54–56
contracting, 104–110; bonuses in, 107; dead work in, 104; defined, 101; labor problems in, 108; miners' feelings about, 103; pay for, 103
contributory negligence doctrine, 140
Cooper, C. A., 113
Copley, D. C., 154
copper mining, 23–24
Corbett, Gentleman Jim, 53
Cornishmen: clannishness of, 5; and contracting, 108–109, 114–115; as highgraders, 120, 122; living standard of, 102; and miners' consumption, 93; and opera, 46; origin of "Cousin Jack" nickname for, 5; as percent of miners, 8; rivalries among, 132, 147; saloons of, 45; superstitions of, 96; and wrestling, 53
corporate paternalism, 26
cosmopolitanism, 20, 22, 23, 46
cost of living, 99–101
Courier (Georgetown, Colo.), 16
courts, 139–140
Cousin Jacks, 96, 115, 165; origin of name, 5. *See also* Cornishmen
Cox, W. Rowland, 86, 94
Coyote Mine, 104
Crampton, Frank: on conditions in mining camps, 11, 19, 22, 27, 30, 43, 58; early life of, 6, 165; friends of, 11, 79, 93; on high-grading, 118, 119, 120; on labor-management issues, 131, 134, 149; mining experiences of, 6, 11, 71–72, 92–93, 96, 108
creditors, 116
Creede, Colo., 21
Cripple Creek, Colo.: cost of ore extraction in, 81; entertainment in, 33, 40, 53; fire department of, 38; high-grading in, 120, 122; influenza epidemic in, 44; juvenile gangs in, 33; labor and management in, 6, 10, 112, 128, 151, 153, 155; living

conditions in, 27, 28, 29, 30, 34, 44, 99; working conditions at, 63, 81
Cripple Creek Times, 38
crosscut, 59
Cross Cut Tunnel, 135
Crown Point mines, 69, 89–90
Curtis, Uri B., 104

Daisy Mine, 115, 165
Daly, Thomas F., 80
dance halls, 21, 22, 24
Davis, Carlyle C., 7
dead work, 68, 104
death benefits, 48–49
Deer Mine, 98
Delamar, Nev., 38, 49–50, 93–94
Democratic Party, 158, 159
Dempsey, Jack, 164
Denver, Colo., 46
Denver and Rio Grande Railroad, 133
depression, 8–9
detectives: as "bad men," 156; as spies on workers, 123, 124, 126, 132, 142–143, 147; union reactions to, 151
Diamond Mine, 116
diarrhea, 41–42
Dickey, J. C., 87, 92
diphtheria, 42, 43, 58
disabled miners, 148
diseases: cerebrospinal meningitis, 42; cholera, 58; colds, 42; diarrhea, 41, 42; dysentery, 41–42, 95; hookworm, 95; influenza, 42, 43–44, 58; miner's consumption, 80–81, 93–94, 142; pneumonia, 42, 43, 93; respiratory ailments, 5, 42, 80–81, 93–94, 95, 142; scarlet fever, 42; silicosis, 93–94; smallpox, 42, 43, 58, 77; typhoid fever, 42, 43, 95; tuberculosis, 5, 42, 93, 95; venereal disease, 30, 42; yellow fever, 42
diseases, causes of, 19, 40, 42
Donovan, Robert E., 134
double-jacking, 54–55, 56, 70
Douglas, Frances (DeKalb), 31
Downing, William, 136
drift, 59
"Driller's Troubles, The," 73–74
drilling: contests of, 54–56 hazards of, 78, 80, 93; techniques of, 54–55; 70–71
drugs, 30
drunkenness, 39, 153, 155
Dubofsky, Melvyn, 147

dust: from drilling and blasting, 71, 80; as health hazard, 5, 93–94; in mines, 19; from quartz, 93–94
dynamite, 71, 78, 79
dysentery, 41–42, 95

Eagles (fraternal organization), 48
earthquake, 91
Eckles, T. H., 157
Edesa, Rosendo, 77
efficiency men, 127
Eilers, Karl, 153
Elder, Mrs. D. H., 31
electricity, 86, 153–154
Elks (fraternal organization), 48
Ely, Nev., 12, 63
embezzlement, 148
emigration, 3–7. See also immigrants
Emmet Guards, 49
Empire, Colo., 10
Empire Mine, 89
Englishmen, 8
equipment, 8
Europeans, 5; in dangerous work, 132–133; as immigrants, 9; prejudice against, 134; in rivalries, 147. See also individual ethnic groups
explosives: as cause of cave-in, 91; hazards from, 78–79, 86–87; on Independence Day, 51; safety measures for, 79; use of, described, 71
extractories, 64; conditions in, 66–68; definition of, 65

Fahnestock, Kenneth L., 81, 84, 128, 129, 143
Fairfield, Nev., 115
falling: of miners, 77–78; of objects, 76
Farrish, Thomas, 138
Faunce, Elmer, 18
Fay, Albert H., 132
Federal Bankruptcy Statute, 137
fellow-servant rule, 139–140
fences, 121, 122
Fifth Judicial District, Colo., 158
Finnish, 9
fire-fighting, 38; as fraternal organization, 48, 49; in mines, 90
fires, 36–39; causes of, 37; compared to snowslide, 16; in mines, 88–91
fishing, 53
floods, 18–19
Florence, Colo., 10
Florence Mine, 118

food, 41
football, 53
Foster, Ernest LeNeve, 114, 141
Francis, G. E., 127–128
Frantz, Albert, 56
fraternal organization, 48, 49
freight charges, 115
French, 45
Friedman, Lawrence M., 139
funerals, 50, 90
fuses, 72, 79

Gallardo, Criso, 79–80
Galloway, James: on climate, 13; death of, 86; expectations of, 3; house of, 29; stock speculation of, 45–46; wanderlust of, 10
Gallup, _____ (contractor), 108
gamblers, 121
gambling dens, 22, 25, 44, 45, 58
Gans, Joe, 53
Gardner, E. D., 130, 134, 141
gases, 69, 71, 86–88
Georgetown, Colo., 10, 14, 60; miners' earnings in, 102, 108–109, 116; ventilation in mines near, 87; winter in, 15, 16–17
Georgia Gulch, Colo., 10
Germans, 46, 132
Giese, W. E., 125
Giles, A. D., 107–108
Gilman Mine, 114
Gilpin County, Colo., 66
giraffes, 66, 85; defined, 67n
Globe, Ariz., 134
Goggin, Ben, 126
gold, 3, 21, 118
Golden, Colo., 18, 120
Goldfield, Nev., 4, 12; assayers in, 120; high-grading in, 118; housing in, 27, 29; strike in, 152; working conditions in mines of, 63, 123
Gold Hill, Nev., 32; guard unit of, 90; mine fires at, 89–90; unions in, 148
Gold Hill Miners' Union, 149, 158
Gold Run, Colo., 10
Goodsprings Camp (Nev.-Calif. border), 19
Goodwin, J. C., 138
Goodwin, R. G., 138
Gould and Curry Mine, 88
Granite Mountain, 90
Grant, Edward, 152
Great Basin, 12–13

Greeks, 5, 133
Gregory Consolidated Mines, 87
Griffith Mountain, 18
Groesbeck, H. V. S., 136
Gruenwalds (operatic stars), 46
Gunnison County, Colo., 10

Hall, Frank, 50–51
Hamlet, 46, 47
Hammond, John Hays, 143
Hanington, C. H., 33, 49, 53
Hanover Mining and Milling Company, 154
hard-rock mining. *See* miners; mines; mining
Harper, John, 138
Harrington, John, 11, 93
Harris, W. R., 112
Harvey, Richard, 5, 136; family of, 5
Haskell, Rachel, 16
Hastings, Edward R., 109
Haywood, William, 150, 151, 157, 160
hazards, mining: double pay for, 73; from dust and gas, 69; from explosions, 59, 72–73; from falling objects, 76–77; from machine drilling, 80; from transportation, 66, 67
hazards, nonmining: daily, 44, 75; winter, 16
Hazlett, R. H., 113
health, 5, 21
health care, 42–43, 142, 150
heat, 88
high-altitude environment, 12–13, 69–70
high-grading, 118–123; attempts to curtail, 153; defined, 104; ore in, 115, 118; reasons for, 121–122, 122n
Hill, Joe, 10
Hill, Nathaniel P., 7
Hillerton, Colo., 28
hoisting engineers, 83–84
Holmes, Wyo., 136
Homestead (bordello), 30
hookworm, 95
horses, 63, 81, 95
horseshoes, 95–96
Hoskins, James, 77
hospitals, 142
housing, 100
humidity, 69
hunting, 53
hydrothermal regions, 66, 69, 87–88

Ibex Mine: accidents at, 75; cages at, 84; high-grading in, 119; naming of, 98; pay for contract work in, 104; shifts in, 63–64
Ibex Mining Company, 106
Idaho Springs, Colo., 15, 102, 128
immigrants, 26, 33, 133–134. *See also* emigration
Independence, Colo., 28, 63
Independence Day, 51–52, 54
Independence Development Co., 138
Indians, American, 9, 22, 26, 96–97
industrialization, 81–82, 168
Industrial Workers of the World, 126, 149
inflation, 127, 149
influenza, 42, 43–44, 58
informers, 123, 124, 125
insurance: employers' liability, 32, 141–142; as expense, 101; lack of, 32, 37; life, 48–49; from union, 145
International Union of Mine, Mill and Smelter Workers, 126. *See also* Western Federation of Miners
Irishmen, 45, 120, 134; as percentage of miners, 8; rivalries of, 132, 147
Irwin, Richard, 10
Isaac, John, 84
Italians, 133–134; and opera, 46; rivalries of, 132; saloons of, 45; translations for, 9
IWW. *See* Industrial Workers of the World

Jackson, W. Turrentine, 50
James, George O., 10
Jefferson County, Colo., 18
Jerome, Ariz., 25, 95
Jesse Knight Mine, 137
Johnson, J. J., 6
Jones, J. J., 75
juvenile gangs, 39

Kearns, Tom, 163
Keena, Leo J., 40; describes prostitutes, 44; on high-grading, 153; on miners' thefts, 124; on mining innovation, 81
Kelley, Pat, 84
Kenney, Leo, 159
Kentuck Mine, 90
Ketchel, Stanley, 53
Kidoo Mine, 98

Kimberly, Utah, 135
Kimberly system, 153
Kindall, Cleve E., 93
King, Alfred Castner, 59, 91
Kinnear, Bertha, 155
Kinnear, J. C., 115, 152, 155, 165
Klais, John and Michael, 110
Knights of Pythias, 48
Knuckey, C. L., 125
Kokomo, Colo., 32

labor: conflict of, with managers, 39,
    116–117, 123, 136–137; laws con-
    cerning, 140; recruitment of, by
    railroads and mines, 5–6; shortages
    of, 5; sympathizers with, 24; turn-
    over of, 10
Labor Day, 48, 52, 54, 125
Lafferty, Edward M., 165
Landers, William, 107
Langrishe, John, 46
lanterns, 63 and n, 68
Lanza, A. J., 93
Larock, S. A., 103, 105
latrines, 95
law and order, 39–40
Lawrence, R. D., 113
laws: needed to protect labor, 136,
    149; miners' recourse to, 137–141
lead miners, 147
Leadville, Colo.: decline of, 23; labor
    relations in, 63, 112, 115, 125, 126,
    147–148, 152, 156, 158; living con-
    ditions in, 27, 101; picnic in, 48
leasing, 110–117; advantages of, to
    owners, 110–111; Cornishmen and,
    114–115; defined, 101–102; expens-
    es of, 115; profits and losses in, 110
Lee, Mabel Barbee, 43; book by, 53;
    on miners' consumption, 94; re-
    members fire, 38; on winter sports,
    52
Lee Basin Mining Company, 107
legal system, 122, 156
legislation: backed by unions, 158–
    160; effect of, on working hours,
    63; to protect miners, 140; union
    demands for enforcement of, 150
level, 59
libraries, 25–26, 47
lightning, 18, 37, 97
Lilian Mine, 114
literary societies, 47
litigation, 137–139, 151

Little Chief Mine, 131
Livermore, Robert, 80, 86, 116, 156
living conditions, 20, 27–30, 100, 167–
    168. See also towns, mining
lockouts, 150
London Guarantee and Accident Co.,
    Ltd., 141
Lookout Mountain, 15
Lord, Eliot, 22; on character of min-
    ers, 73–74; on Crown Point funeral,
    90; on labor relations, 132, 157; on
    mining accidents, 83, 88, 141; on
    mining hazards, 69, 88; on religion,
    164
Lowell, Mass., 23

Mahala Mining Claim, 113
Mahany, J. G., 136
management, 149; attitudes of, toward
    miners, 123, 142–144; deception by,
    5; difficulty of, in meeting payroll,
    135; and high-grading, 121–123;
    and mines, 130–136; opposition of,
    to unions, 124, 128, 129, 150–151
Manhattan, Colo., 75
Manhire, Rue and Charles, 110
Manogue, Father Patrick, 164
Manzanita Mining Co., 106–107
Mapes, Dorchester, 143
Marchetti, Celeste, 77
Marlowe, 46
martial law, 156
Mason, Henry, 84
Masons (fraternal organization), 48
Mayfield, Henry, 74
McGrath, James, 136
McKay, Neil, 110
mediation, 154, 156
merchants, 121
Mexican-Americans: feast days of,
    50; lack of acculturation of, 26;
    prejudices against, 9, 119, 134, 147;
    superstitions of, 96–97
Middlemarch Mine, 143
militia, state, 156
mineowners associations: assessments
    by, for detectives, 125–126; at-
    tempt to reduce wages, 127–129; in
    Cripple Creek, 130
miners, 11–12, 58; attitudes of, 3–4,
    23, 92–93; 144, 163, 168; back-
    grounds of, 3–5; character of, 11,
    58; comradeship among, 92–93;
    consequences of illness for, 42–43;
    dangers to, 78 (see also hazards,

mining); diversions of, 25–26, 44–57, 58, 73–74; equipment of, 7–8, 72; high-grading among, 118–120; labor relations of, 99, 101, 119, 126, 128–129, 135–139, 144; living conditions of 20, 27–30, 100, 167–168; in mine fires, 36–37, 89, 90; as pioneers, 161; poetry about, 59, 73–74, 75, 91, 145; shifts worked by, 15, 60, 63–64, 102, 149–150; superstitions of, 95–96, 97–98; tasks of, 68, 70–74; wages of, 99, 101; wanderlust of, 10; work clothes of, 60 and n, 66, 119; working conditions of, 63, 66. *See also* management, attitudes of, toward miners

miners' consumption, 80–81, 93–94, 142

Miners' Magazine, 47, 156; as official WFM organ, 24–25; on artificial wages, 127

mines: communication in, 67; consolidation of, 64–65; fires in, 87–88; mechanization of, 65, 81–82, 86, 139, 142, 166–167; size of, 64–65, 82; transportation in, 66–68

mining: contracting in, 72, 101, 104–110; effect of winter on, 14–15; leasing in, 101, 110–117; recruiting labor for, 5; safety in, 82, 86, 129, 130, 139–140 (*see also* hazards, mining); specialized tasks in, 64–65, 68, 71, 73; shutdown of, from fire, 37; terminology of, 59–60; wage work in, 101

Mining and Scientific Press, 25, 144

Missouri City, Colo., 34

mobility: geographic, 3–7, 10, 161–162; social and economic, 162–166

Moffat, David H., 113, 123

Mogollon, N.Mex., 43

Mogollon's Deep Down Mine, 88

Mohawk Mine, 118

Montana House of Representatives, 164

Morgan, Cecil C., 104, 135

Mormons, 45

Morris, Henry C., 99, 115

Moudy, Ross B., 10, 87, 112, 164

Moyer, Charles, 148, 150, 157

mucking, 70

Mudd, S. W., 125, 153

mudslides, 18

mules, 52, 95, 96

Murphy, John H., 89, 157

National Association of Manufacturers, 167

National Council for Industrial Defense, 167

National Mine, 119

Nevada, 11, 12, 13

Nevada City, Colo., 34

New Mexico, 11, 12, 13

newspapers, 47, 134

New Year's, 50

New York Call, 24, 134

Northey, Sam, 94

Nylund, _____ (union secretary), 148

Oates, Joseph, 77

Oberto, Steve, 127–128

O'Brien family, 29. *See also* Colwell, Pearl O'Brien

occupations, nonmining, 10, 22

Odd Fellows, 48

Oddie, Tasker, 115

Ogden Standard, 134

O'Gorman, M., 143

Olcott, Eben, 15, 18, 26

Old, Robert Orchard, 18, 64

Old Dominion Copper Company, 134

Olvany, John, 113

Ontario Mine, 88

Open Sesame Mine, 106

opera, 22, 46

Ophir Mine, 88, 97

Orchard, Harry (Albert E. Horsley), 157

ordinary risks, rule of, 139

ore, 115; buckets for, 67; cages for, 67–68; exhaustion of, 36

Orientals, 9, 22, 30, 33, 159

Our American Cousin, 46

Owers, Frank W., 158

Panic of 1893, 123, 162, 166

Papanikolas, Helen Zeese, 133

Park City, Utah, 13

Parker, J. D., 130, 141

Parsons, George Whitwell: attitude of, toward miners, 74; background of, 8, 165; and contract mining, 108; decision of, to mine, 4, 8; experiences of, in mining communities, 18–19, 28, 42, 63; on mining, 65, 71, 78; observations by, con-

cerning mining communities, 14, 18–19, 41.
Paul, John Marshall, 110
Pearce, Richard, 78
Pelican Mine, 78
Penrose, William J., 164
Perrigo, Lynn, 102
personal injury suits, 139–141
pests, 40–41
Pettibone, George, 157
Pevine, Nev., 84
Phelps Dodge Company, 10, 26, 105
picket lines, 7
Pierce, George, 38
Pima County, Ariz., 104, 122
Pinkerton National Detective Agency, 119, 125
Pinon Mine, 106
plagues, seasonal, 42
plumbing, 25; lack of, 29
pneumonia, 42, 43, 93
politics, 26, 157, 158; in WFM, 159–160
pollution, 21
Pomeroy, Richard A., 14, 16
population shifts, 160–161
Populists, 158, 159
Possauer, Mrs. George, 14
Potosi Mining Co., 42
practical jokes, 25–26, 56–58, 97, 163
precious minerals, 3–4, 21
Prescott, Ariz., 51
Price, B. A., 73
Pride of the West Mining and Milling Company, 142
Prince, L. Bradford, 103, 107, 108, 113; contracts of, 105–106
profit motive, 167
prostitution, 19, 29–31, 44; red light districts of, 22, 25; reason for turning to, 33
Protestants, 164
pumping, 88

racing, 53
Ragan, Matt, 75
railroads, 5, 49, 99
Rambler Copper and Platinum Company, 143
Randall, Jesse S., 16
Rawhide, Nev., 4
Reindeer Mine, 134
religion: Catholic, 164; Mormon, 45; Protestant, 164
Republicans, 159

Requa, Isaac, 42
respiratory ailments, 5, 42, 80–81, 93–94, 95, 142; colds, 5; miners' consumption, 80–81, 93–94, 142; pneumonia, 42, 43, 93; silicosis, 93–94; tuberculosis, 5, 42, 93, 95
Richards, C. O., 110
Rickard, Thomas A., 118, 119, 120, 121–122
Rico, Colo., 31
Ridenour, G. J., 138
Roadside Mine, 31
rock-drilling contests, 52, 54–56
Rockies, lower: environment of, 12–13
roller-skating, 52
Rosita, Colo., 10
Rossa, Joe, 105–106
royalties, 110, 113–115
Ruby Hill Mining News, 164
rushes, 3–4, 21
rustling cards, 129, 152n

safety, 60n, 77, 80
"Safety Organizations in Arizona Copper Mines," 141
St. Elmo, Colo., 99
St. John, Vincent, 129
St. Lawrence Mining Co., 135–136
St. Patrick's Day, 50
Salado, Jesus, 85
saloons, 22, 25, 44; licensing of, 39; as receivers of high-grade, 121; as social centers, 45, 58
Salt Lake City Herald-Republican, 153
Sanders, Thomas D., 3–4
Sanford, Albert Byron, 18, 28, 76, 109; father of, 28
sanitation, lack of, 21, 40, 95
San Juan Mountains, 7, 10
Savage Mine, 33
Sayre, Elizabeth, 32
scabs, 151, 156
scarlet fever, 42
schools, 22, 34, 43
Schuler, Mrs. _____, 18
scrip, 135
Seven-thirty Mine, 60
shaft, mine, 59
Shakespeare, 46, 47
Sherman, Gerald, 105
Sherman Silver Purchase Act, 162
shifts, 149–150; continuous, 60; eight-hour, 63, length of, 15, 63–64, 102

shivaree, 56
Shovin, Edward, 114
Shumer, John, 13, 104
shutdowns, seasonal, 15
silicosis, 93–94. *See also* miners' consumption
silver, 3, 21, 118
Silver City, Idaho, 148
Silver Plume, 60
Silverton, Colo.: blizzard in, 15; influenza epidemic in, 44; leasing in, 113, 115–116; winter in, 16
Silver Wing Lode, 135
single-jacking, 54–55, 56, 70
Siringo, Charlie, 119, 124, 125, 132
Skliris, Leonidas G., 133–134; and three brothers, 133
Slavic, 9
Slayback Mine, 111
Sleeping Beauty Mine, 106
smallpox, 42, 43, 58, 77
smells: of towns, 20–21; of offal, 40; in mines, 69, 95
smelter fumes, 21
Smith, B. P., 160
Smith, Duane A., 9, 131, 162
Snake River, 42
snipe hunts, 57
Socialism, 158, 160
Socialist Party of America, 160
Socorro Mining and Milling Co., 79, 86, 88
South Pass, Wyo., 4
Spanish Bar, Colo., 50
speculation, 45–46
spiritualists, 97
spontaneous combustion, 37, 89
sports, 53. *See also* individual events
Sprague, Marshall, 120
stamp mills, 21
Star District, Utah, 13, 104
Starne, Maurice, 113
"Star Spangled Banner, The," 51
statutory negligence, 140–141
Staunton, W. F., 14, 40
step devil, 96–97
Stetson Hat Company, 149
Stevens, William H., 135
Stevens Mining Company, 69–70, 87, 102
Stewart, George H., 76
Stivens, Thomas, 78
stockholders, 130
stock speculation, 45–46
stope, 59

stores, 22, 25
strikebreakers, 6, 128, 147, 155–156
strikes, labor: in Bingham Canyon, 133–134; causes of, 123, 150, 153–154; community attitudes toward, 155; effects of, on leasers, 115–116; of 1893, 6; in Leadville, 126, 131; mediation of, 154, 156; as political action, 157–160; sympathy, 156–157
strikes, ore, 3, 161
Struber, George, 10
Sullivan, James A., 84
Sullivan, John L., 53
Sullivan, Michael, 11, 79, 93
sulphides, 153
Sun and Moon mines, 128
Sunset Mine, 78–79
superstitions, 95–96, 97–98
Sutro Tunnels, 88
Swedes, 10, 120

Tanner, Charles H., 127
Tarryall, Colo., 10
technological change, 166–167; as cause of increase in accidents, 81–82; personal injuries from, 86, 139, 142
Teller County, 50
Telluride, Colo., 34, 148
temperature: extremes of, 12–13, 19; in mines, 69–70
tennis, 53
tents, 27
Thanksgiving Day, 50
theater, 22, 46–47
theft: of mining supplies, 122–123, 124; of ore, 123; precautions against, 66; of union funds, 148. *See also* high-grading
Thiel Detective Agency, 124–125
Thomas, Lowell, 53
three-hundred-day year, 102n–103n
timbering: as accident cause, 77, 78; as cause of cave-in, 91, 92; in contract work, 72; in fires, 89, 90
Todd, Arthur, 103
Tombstone: cost of living in, 99; eight-hour day in, 63; railroad celebration in, 49; shacks in, 28; wages in, 4
tommyknockers, 96
Tonopah, Nev., 4; boom in, 12; consumption in, 93; drilling contest in, 56; eight-hour day in, 63; fire

in, 37; housing in, 27; railroad days in, 54

towns, mining, 20–27; company towns as, 23–26, 131–132; cosmopolitanism of, 20, 22, 23, 46; lack of urban planning in, 20; urban character of, 166, 167–168. See also living conditions

tramming, 70, 85

transportation: of freight, 99; in mines, 63, 66–68; public, 63; to work, 63, 75

traveling card, 152, 152n

Traylor, John, 157

trees, 16

Trembath, John, 157

Truscott, William, 84

tuberculosis, 5, 42, 93, 95. See also miners' consumption

tunnels, 106

Turquesa, N.Mex., 103

typhoid fever, 42, 43, 95

unemployment, 10, 15

Union Mine, 86

unions: conflict within, 146–148; cooperation within, 148–149; job assistance from, 149; management feelings toward, 124; political action by, 157–160; reasons for forming, 145–146, 149–151; social events of, 52, 146; strikes of, 151–157; violence in, 155n, 157. See also Western Federation of Miners; Industrial Workers of the World

United States Bureau of Mines: accident statistics of, 77, 82, 102n–103n; advice from, 78, 93, 95; mine-fire statistics of, 89

United States Census Bureau: ethnic statistics reported by, 8, 9n; statistical irregularities in, 11; report by, of number of miners in intermountain west, 1870–1920, 11–13

United States Reduction and Refining Company, 126

United Verde Mine, 95

urbanization, 23, 166, 167–168

Utah, 11, 12, 13

Utah Copper Company, 24, 133

Vaneck brothers, 123

Van Wagenen, Theodore F., 143–144

venereal disease, 30, 42

ventilation: lack of, in mines, 69, 86–88, 95; systems of, 87

Venture Corporation, 153

Victor, Colo., 63

Victor Club, 40

violence, 40, 155n, 157

Virginia City, Nev.: accidents in, 37, 38–39, 84, 88, 89; business in, 22, 45, 127, 151; climate of, 13, 15–16; history of, 8, 22, 23; juvenile gangs in, 33; living conditions in, 13, 15–16, 28, 34, 37, 38–39, 47; mining in, 63, 89; population of, 8; railroads in, 49; unions in, 148

Virginia City and Truckee Railroad, 49

Virginia City Enterprise, 127

Virginia City Miners' Union, 90, 148, 158

Vulture Mine, 115

wages: changes in, 99; compared, 4, 102; in Leadville, 101, 127; liens for, 137–139; reductions of, by owners, 127–128; scales of, 102–103, 149; of typical miner, 102; in "wage work," 101

Waite, Davis H., 156

Waldorf, Adeline, 97

Waldorf, Jacob, 7, 158

Waldorf, John T.: and Hamlet, 47; on Independence Day, 51; in juvenile gang, 33; and library, 47; and memories of fire, 38

Walker, Arthur L., 143

War Dance Mine, 110

Ward, Artemus, 47

Warman, Si, 21

Washington's Birthday, 50

Washoe zephyrs, 17

water, 69, 88, 91

weather, 12–19, 91

Welsh, 132

west, intermountain: beauty of, 48; defined, 3; number of miners in, 11–13

Western Federation of Miners (WFM): as agent of change, 154; attitudes of, toward immigrants, 9, 125, 147; boycott by, 149; compared with other unions, 167; conventions of, 145, 146, 148, 149, 159; formation of, 123; labor actions by, 63, 136–137, 149, 152, 156–157; and Labor Day, 52; magazine of, 24–25; organizational efforts of, 24, 123, 125; policies of, 63, 112–113, 127, 128, 148, 149, 151, 159–160; relations of, with management, 126, 143

Western Pacific Railroad, 133

WFM. *See* Western Federation of Miners

White, Joseph H., 95

White Pine District, Nev., 8, 15, 50, 87

Whitney, Sumner, 32; widow of, 32

Wickenburg, Henry, 115

Wilcox, Ariz., 134

Williams, Albert, Jr., 78

Williams, Ben, 26

Williams, Dick, 40

William Shea & Company, 120–121

Winnemucca, Nev., 119

Witherell, W. Frank, 135

Wobbly, 10. *See* Industrial Workers of the World

Wolftone Consolidated Mining and Milling Company, 113

Wolgast, Ad, 53

women, 29–34; daily life of, 16; and fire disaster, 89, 90; and houses, 29; lack of, 26; as miners, 31; occupations of, 32; as prostitutes, 29–31, 44; rights of, 31; superstitions about, 96; WFM auxiliary for, 146 and n; as widows, 31, 32, 140

Wood, Henry E., 27, 79, 114, 116

working days, 102n–103n

wrestling, 53

Wright, William, 17, 82, 90, 97

Wyman, Walker D., Jr., 81

Wyoming, 11, 12, 13–14

yellow fever, 42

Yellow Jacket Mine, 85, 89–90

Yenkes, Thomas M., 104–105